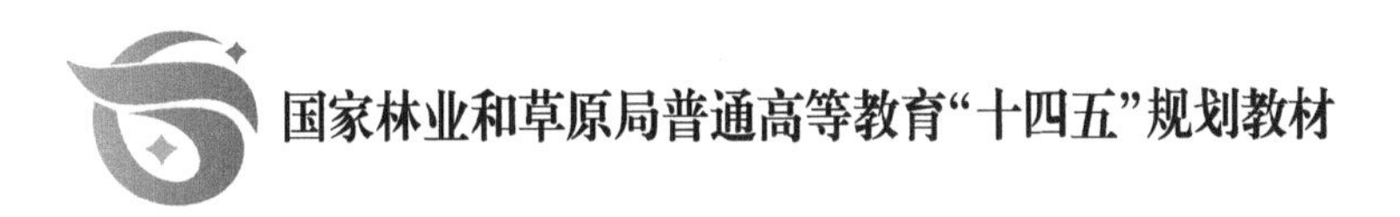

地理信息系统原理

Fundamental of Geographic Information System

张青峰　主编

中国林業出版社
CFPH China Forestry Publishing House

图书在版编目(CIP)数据

地理信息系统原理 = Fundamental of Geographic Information System / 张青峰主编. —北京：中国林业出版社，2022.8

ISBN 978-7-5219-1783-3

Ⅰ.①地… Ⅱ.①张… Ⅲ.①地理信息系统 Ⅳ.①P208

中国版本图书馆 CIP 数据核字(2022)第 130478 号

中国林业出版社教育分社

策划、责任编辑：范立鹏　　**责任校对**：苏　梅

电　话：(010)83143626　　**传　真**：(010)83143516

出版发行　中国林业出版社(100009　北京市西城区刘海胡同 7 号)
E-mail：jiaocaipublic@163.com
http://www.forestry.gov.cn/lycb.html

经　销　新华书店
印　刷　北京中科印刷有限公司
版　次　2022 年 8 月第 1 版
印　次　2022 年 8 月第 1 次
开　本　787mm×1092mm　1/16
印　张　14.375
字　数　340 千字
定　价　45.00 元

《地理信息系统原理》编写人员

主　　编： 张青峰

副主编： 刘　京　成　遣　何红梅

编　　委：（按姓氏笔画排序）

王　琤（西北农林科技大学）
王新军（新疆农业大学）
成　遣（沈阳农业大学）
刘　京（西北农林科技大学）
刘金成（西北农林科技大学）
何红梅（西北农林科技大学）
张廷龙（西北农林科技大学）
张青峰（西北农林科技大学）
张楚天（西北农林科技大学）
郑子成（四川农业大学）
庞国伟（西北大学）
周淑琴（山西农业大学）
晋　蓓（西北农林科技大学）
彭强勇（四川农业大学）

前　言

随着我国北斗卫星导航系统建设的蓬勃发展，地理信息系统(GIS)技术在测绘地理信息、耕地保护、自然保护地监管、地质矿产、海洋事务、国土空间规划、生态保护修复、灾害预警防范、调查监测、林草碳汇计量等领域的应用日趋深入，使得该技术本身在普及应用中也得到了飞速的发展。

“地理信息系统原理”是地理信息科学相关专业的一门基础课和必修课，旨在使学生理解和掌握地理信息系统的基础知识、基本理论、基本方法和地理信息系统软件的基本操作技能。该课程的教学内容包括地理信息系统基本概念、地理信息系统软件介绍、现实世界的表达、地理数据的基本属性、地理数据建模、地理数据库的建立与维护、地理信息系统空间分析、地理信息系统决策应用等，形成一套完整的数据采集、入库、分析、表达、应用的教学内容体系。通过本课程的学习，可为学生正确理解、应用和表达地理数据打下方法论基础。

自 2000 年以来，我国各高等院校陆续开设了地理信息系统相关的双语和英文教学课程，编写出版与之配套的教材是该类课程建设的重要方面。为此，按照教学大纲要求，本教材编写人员立足自身教学科研实践，根据现阶段地理信息系统技术发展情况，系统总结和借鉴地理信息系统领域国内外优秀教学科研成果，从科学性、系统性、逻辑性和实用性方面考虑编写了本书。本书内容包括五个基本版块：概述(第 1 章和第 2 章)、原理(第 3 章和第 4 章)、技术(第 5 章和第 6 章)、分析(第 7~10 章)和应用(第 11 章和第 12 章)等，介绍了若干地理信息系统应用案例，以反映学科最新实践。

本书由张青峰担任主编，刘京、成遣和何红梅担任副主编，编写具体分工如下：绪论、第 1 章由张青峰编写；第 2 章由周淑琴和刘金成编写；第 3 章和第 5 章由刘京和张廷龙编写；第 4 章由张青峰和张楚天编写，第 6 章和第 10 章由成遣和晋蓓编写；第 7 章由郑子成和彭强勇编写；第 8 章由庞国伟编写；第 9 章由王琤和张青峰编写；第 11 章和第 12 章由王新军和张楚天编写。全书最后由张青峰统稿，刘京、成遣和何红梅共同审定。

本书的编写和出版得到西北农林科技大学、西北大学、沈阳农业大学、四川农业大学、山西农业大学、新疆农业大学，以及国内外诸多专家与同行的帮助与支持，在此谨向他们表示最诚挚的感谢！

当前，随着相关学科理论研究和实践的进步，地理信息系统技术在理论和应用方面也不断地更新和发展，本书不可避免存在一些疏漏和错误，我们真诚地欢迎广大读者和各位同仁批评、指正。

编　者

2021 年 12 月

前言

Preface

As Geographic Information System (GIS) is becoming more and more important and widely used, this discipline is making our daily study on geography easier, more efficient and enjoyable. With the advent of faster, more powerful computers with increased storage potential in the past decade, the rapid development of GIS and spread from mainframe computers to mobile terminal equipment via minicomputers, workstations and personal computers has brought a powerful capability to a much larger community of users. New GIS-related technologies are constantly being generated and applied to a wide range of work and life. As a science, a technology, a discipline, and an applied science for problem solving, GIS is of enduring importance because of its core principles, its applications, its unique analysis methods, and the host of practices and dilemmas that often accompany the use and communication of geographic information.

With the popularity of GIS education since the year of 2000, teaching reform on bilingual or English GIS course teaching in major universities has gradually been implemented in China. Thereby, suitable textbooks have long been a must. However, it was a challenging task to work out suitable and high-quality textbooks for GIS teaching under such limitations as credit hours, related resources and etc. In our long-term teaching activities, a variety of referenced GIS books and materials, combined with the results of scientific research projects, have been gradually collected, classified and arranged. The textbook *Geographic Information System* (GIS) is organized with the purpose of encouraging the development of the bilingual/English GIS teaching. It covers the basic GIS theories, technologies, analysis and applications in a comparatively rounded system. Referring to abundant foreign literature and resources has made the book more systematic and objective.

The textbook is organized in 5 major but interlocking sections. Each section investigates the unique, complex, and difficult problems that are posed by geographic information, and together they build into a holistic understanding of all that is important for GIS. The first section provides a general introduction in Chapter 1 followed by GIS software in Chapter 2. The following sections focus on principles (Chapters 3 and 4), technologies (Chapters 5 and 6), analysis (Chapters 7, 8, 9 and 10) and Application (Chapters 11 and 12). In Chapter 1, the concept of GIS, GIS-related disciplines, history, components and applications are introduced. In Chapter 2, the development process, architecture, basic functions, types and major producers of GIS software are described. In Chapter 3, GIS models, including raster data model and vector data model are introduced. In Chapter 4, the concept of geographical phenomena, the classification of spatial objects, spatial autocorrelation, spatial sampling methods, spatial interpolation and uncertainty are introduced. In Chapter 5, the concept of data models such as vector data model and raster data model are introduced. In Chapter 6, the establishment and maintenance of Geodatabase is briefly introduced. In Chapter 7, basic spatial analysis function-geographic query and spatial measurements is introduced. In Chapter 8, features and processes of vector-based spatial analysis and raster data-based spatial analysis are introduced. In Chapter 9, the generation, operation, interpolation, expression and application of digital terrain models are introduced. In Chapter 10, the concept, basic elements and common functions of network are introduced. In Chapter 11, the concepts, methods and applications of GIS decision support function are introduced. In Chapter 12, the applications of GIS in different areas are briefly listed. In this course, students will not only learn specialized knowledge on the basic principles of GIS, including the concepts and the techniques of handling geographical data, the skills and techniques to input, manage, analyze and display spatial information, the concepts and techniques for spatial data analysis and modeling, but also acquire major-related linguistic proficiency in English.

The first edition of this textbook has been used for more than five years. Due to the teaching reform and knowledge updating, the teaching content needs to be renewed. According to the requirements of the syllabus, the basic framework of the first edition has been remained, but the specific contents have been revised. The relevant data, basis and contents are updated. Considering the scientificity, systematicness, logicality and practicability of the teaching materials, a few paragraphs are added or re-organized so as to reflect the latest practice of the discipline.

The revised textbook is writed by highly experienced teachers and researchers from Northwest A&F University, Northwest University, Shenyang Agricultural University, Sichuan Agricultural University, Shanxi Agricultural University and Xinjiang Agricultural University. We are responsible for the content in the book and many of the materials contained are shared and contributed by friends and professional peers in and outside the GIS academic community. We greatly appreciate their invaluable support and help from many organizations and individuals that have made different academic resources and words into a complete edition.

With the constant updating and development of GIS technology due to the theoretical research content and practical requirements of GIS-related projects, there are inevitably some flaws and even mistakes in this revised textbook. Your kind suggestions will be encouraged and appreciated. Hopefully, this book can be a useful resource for your learning or teaching.

Authors

Dec. 02, 2021

Contents

Chapter 1

Introduction to GIS

A geographic information system (GIS) integrates hardware, software, data, network, procedures and personnel for capturing, managing, analyzing, and displaying all forms of geographically referenced information. It enables us to observe, understand, question, interpret and visualize data in many ways and reveal relationships, patterns, current status and tendencies in the form of maps, globes, reports, and charts. Besides, it serves us to resolve queries or solve problems by counting on our data and the easy-to-understand and easy-to-share methods(地理信息系统整合硬件、软件、数据、网络、程序和人员用以捕捉、管理、分析和显示地理参考信息的各种形式。GIS 使我们能够用多种途径来观察、理解、询问、解译和表达数据，并以地图、球体、报告和图表的形式来揭示数据之间的关系、模式、现状和趋势。同时，GIS 有助于我们通过一种便于理解、易于共享的方法，应用地理数据来回答问题或解决问题).

1.1 Geography

Firstly mentioned in the *Book of Changes* in the 5th century BC, the term *Geography* referred to the topographic studies and on-site inspection (地理一词最早出现在公元前 5 世纪的《周易》一书，指的是研究地形和考察环境). Since then, geography (alternative name: *Feng Shui or geomantic omen*, 别称：风水或堪舆) has become a *valiant attempt* (勇敢的尝试) to understand everything that exists or happens on the earth's surface. It is in fact a study of the environment and has close associations with *Astrology* (占星学), *Astronomy* (天文学), Geography, Physics, Mathematics, Biology and Medicine. Therefore, many geographers with the foundation of geology, geophysics, economics, sociology, *anthropology*, *philosophy*, physics, chemistry, or medical science have created extensive spatial analyses. They know a lot about things, and *exploit* (开拓) all about things in detail.

However, the concept of Geography itself is too ambiguous(地理一词的概念本身是模糊不清的). It may have different meanings in different branches of science and technology, and different understandings about the concept of geography exist.

Geography is the study of the earth's landscapes, people, places and environments. *Literally* (从字面意义上讲), the word geography may be interpreted in terms of its component parts: Geo and geography. Geo refers to the earth (geology, earth's crust, land, underground, region, terra, field, location, topographic forms and features, etc.; 地质、地壳、陆地、地下、地区、土地、田地、地点、地形地貌等), and *graphy* means a process of describing (arrangement, reason, mechanism, texture, etc.; 条理、理由、机理、纹理等); thus a literal definition about geography would be to describe about the earth(Derived from the ancient Greek geography of the 2nd century BC, it is three centuries later than China; 来源于公元前 2 世纪的古希腊语"geographia", 比我国晚了 3 个世纪).

Another main definition of geography focuses on spatial relationships. As an academic subject, geography teaches students to understand these spatial (or non spatial) relationships between spatially distributed objects, i.e. *precipitation* (降水), air temperature, population, … A primary tool in studying these spatial relationships is the map because it gives a *vivid portrait* (生动写照) of spatial relationships and phenomena over the earth, whether a small segment of the earth's surface or the *solid world* (现实世界).

Geography has likewise been called the world discipline and it is unique in *bridging* (桥接) the social sciences (*human geography*, 人文地理) and the natural sciences (*physical geography*, 自然地理). Human geography concerns the study of patterns and processes that shape the human society. It *encompasses* (包括) the human, political, cultural, social, and economic aspects. While, physical geography (or physiography) focuses on geography as an earth science. It aims to understand the physical problems and the issues of *atmosphere* (大气圈), *biosphere* (生物圈), *pedosphere* (土壤圈), *hydrosphere* (水圈), and *lithosphere* (岩石圈).

It is the well-known fact that the mankind has fostered, enriched and developed different kinds of civilization in the long history of the human society. Almost all events, activities and things associated with human beings happen or exist on the surface or near-surface of the earth. So, it is truly helpful for us to know where something happens, in daily life and work. Meantime, almost all human activities and decisions involve geographic elements and have geographic consequences. For instance, denudation, grassland degradation and undue cultivation of land can directly contribute to the forming of the ecologically frail areas, soil erosion and the poverty of the people (在人类社会的历史长河中,人类创造、丰富和发展了各种文明。几乎所有与人类相关的事件、活动和事情均发生或存在于地表或近地表。因此,当我们在日常生活和工作中遇到某种情况时,知道事情发生的地点对我们非常有帮助。同时,几乎所有的人类活动和决策都包含着地理要素且均产生地理方面的影响,如森林过伐、草地过牧、土地过度垦殖能直接导致生态脆弱地区、土壤侵蚀和贫困人口的形成). Therefore, it is taken for granted that geographic location is a critically important attribute of events, activities, and things. Geography can be further considered as the discipline related to geographic location(地理进而被认为是与地理位置相关的学科). Issues concerning geographic location can be termed *geographic issues* (地理问题), such as: where to locate public facilities (schools, hospitals, police stations or fire station, etc.) for the governments, where to build new highways that solve the most critical bottleneck of the road network for the *Transportation Authorities* (交通局), where to build new stores or *good distribution centers* (货物配送中心) that provide the most cost-effective services for commercial groups or *individuals* (个人), what is the best route from classroom to the *cafeteria* (食堂) for students, and so on.

In general, all these different conceptual meanings emphasize the study of the physical and human features of the earth, focusing especially on explaining the location of these features and the environmental relationships among them. However, an exactly concept in geography is not universally agreed-upon.

1.2 Information System

Information systems help us to manage what we know, by making it easy to organize and store, access and retrieve, manipulate and synthesize information, and apply it to the solution of problems(Longley et al., 2001) (信息系统有助于我们管理所知道的东西，并使得信息的组织、存储、获取与检索、操作与合成以及应用信息系统来解决问题变得容易).

A system is a group of connected entities and activities which interact for a common purpose, such as the human body's digestive system, respiratory system (系统是一组相互连接的实体和行为，为了共同目的而彼此交互，如人体的消化系统和呼吸系统). An information system is a collection of procedures in which data-operating chains generated from raw data are useful for a wide scope, such as observation, measurement, description, explanation, forecasting and decision-making (信息系统是许多程序的集合，在原始数据上生成数据的操作链，这个操作链应用范围广，如观察、测量、描述、解释、预测和决策制定等). However, data is not equivalent to information in spite of their frequent applications in the GIS arena (数据不等同于信息，尽管它们频繁地应用于 GIS 领域).

①Data (the plural of datum) is raw geographic fact. It has no significance beyond its own existence. In other words, data has no meaning of itself and we can know nothing from it. Data is generally *neutral* (中立的) and almost context-free, and it reveals nothing about its identity or its relationship with other objects. Data can exist in any forms such as numbers, text, or symbols. When digital data are transmitted electronically, they are treated as *bitstream in binary form* (以二进制形式表示的位流): A bitstream is a time series or sequence of bits and a *bytestream* (字节流) is a series of bytes which is typically of 8 bits each, and can be regarded as a special case of a bitstream. Data are assembled together in a database ranging from a Kilobyte to Doggabyte (Table 1-1).

Table 1-1 Potential GIS Database Volumes

Computer Disk Storage	Processor or Virtual Storage (Strictly, bytes are counted in powers of 2)
1 Bit = Binary Digit (0 or 1)	1 Bit = Binary Digit (0 or 1)
1 Byte = 8 Bits	1 Byte = 8 Bits
1 Kilobyte (KB, 千字节) = 1000 Byte	1 KB = 1024 Byte = 2^{10} Byte
1 Megabyte (MB, 兆字节) = 1000 KB = 1 ×10^6 Byte	1 MB = 1024 KB = 2^{20} Byte
1 Gigabyte (GB, 吉字节) = 1000 MB = 1 ×10^9 Byte	1 GB = 1024 MB = 2^{30} Byte
1 Terabyte (TB, 太字节) = 1000 GB = 1 ×10^{12} Byte	1 TB = 1024 GB = 2^{40} Byte
1 Petabyte (PB, 拍字节) = 1000 TB = 1 ×10^{15} Byte	1 PB = 1024 TB = 2^{50} Byte
1 Exabyte (EB, 艾字节) = 1000 PB = 1 ×10^{18} Byte	1 EB = 1024 PB = 2^{60} Byte
1 Zettabyte (ZB, 泽字节) = 1000 EB = 1 ×10^{21} Byte	1 ZB = 1024 EB = 2^{70} Byte
1 Yottabyte (YB, 尧字节) = 1000 ZB = 1 ×10^{24} Byte	1 YB = 1024 ZB = 2^{80} Byte
1 Brontobyte (BB, 波字节) = 1000 YB = 1 ×10^{27} Byte	1 BB = 1024 YB = 2^{90} Byte
1 Nonabyte (NB, 诺字节) = 1000 BB = 1 ×10^{30} Byte	1 NB = 1024 BB = 2^{100} Byte
1 Doggabyte (DB, 刀字节) = 1000 NB = 1 ×10^{33} Byte	DB = 1024 NB = 2^{110} Byte

②Information is meaningful data serving some purposes, or the data that has been given some degree of interpretation (信息是服务于某种目的的有意义的数据，或者是被赋予某种特定解释的数据). It is also the content of a database assembled from raw facts by way of relational connection.

③Evidence refers to results of many data sets based on meta-analyses, which is regarded as a multiplicity of information from various sources.

④Knowledge is the information that is understandable personally, highly processed based on specific context, experience, and purpose.

⑤Wisdom refers to policies developed and accepted, used in the context of decisions made or advice given. Also, it is philosophy-based knowledge impossible to share with everyone (Table 1-2).

Table 1-2 A Ranking of the Support Infrastructure for Decision Making

Decision-making Support Infrastructure	Ease of Sharing with Everyone	GIS Example	Personal Cognition
Wisdom	Impossible	Policies developed and accepted, used in the context of decisions made or advice given	Know why
Knowledge	Difficult, especially tacit knowledge	Personal knowledge about places andissues	Know how
Evidence	Often not easy	Results of GIS analysis of manydatasets or scenarios	Know what
Information	Easy	Contents of a database assembledfrom raw facts	
Data	Easy	Raw geographic facts	Know nothing

1.3 Definitions of GIS

1.3.1 Basic Concept

A GIS is a specific form of information system based on geographical data. In a GIS, information is spatially characterized. It is also a computer-based information system that can capture, model, manipulate, retrieve, analyse and represent spatially referenced data (GIS 是一种特别的基于计算机的，能够获取、建模、操作、检索、分析和表达空间参考数据的信息系统).

Geographical data(地理数据): Spatially referenced data.

Spatial Reference(空间参考): A coordinate, *textual description*(文字说明) or *codified name*(编码) by which information or entity can be related to a specific position or location in space especially on the earth's surface. Spatial reference system is a reference framework consisting of a set of points, lines, surfaces and a set of rules to define the positions of points in space in either two or three dimensions.

Spatial Data(空间数据): Also known as geospatial data or geographic information: data or information that identifies the geographic location of features and boundaries on earth. It is normally stored as coordinates and *topology*(拓扑), and it can be mapped.

1.3.2 Selected Definitions of GIS

Countless people offer definitions of GIS. These different definitions are placed on various aspects of GIS and have been suggested over the years, although none of them is entirely satisfactory. Nevertheless, all include the essential features of spatial references and data analysis. Some definitions are particularly helpful, for example, GIS is/are:

①A powerful set of tools for collecting, storing, retrieving, transforming and displaying spatial data from the real world (Burrough, 1986).

②A system for capturing, storing, checking, integrating, manipulating, analyzing and displaying data which are spatially referenced to the Earth (Chorley, 1988).

③The telescope, the microscope, the computer, and the xerox machine of regional analysis and synthesis of spatial data (Abler, 1988).

④Any manual or computer-based set of procedures used to store and manipulate geographically referenced data (Aronoff, 1989).

⑤An integrated package for the input, storage, analysis, and output of spatial information... analysis being the most significant (Gaile et al., 1989).

⑥An information system designed to work with data referenced by spatial or geographic co-ordinates. In other words, a GIS is both a database system with specific capabilities for spatially-referenced data, as well as a set of operations for working [analysis] with the data (Star et al.,

1990).

⑦Automated systems for the capture, storage, retrieval, analysis, and display of spatial data (Clarke, 1990).

⑧Tools that allow for the processing of spatial data into information, generally information tied explicitly to, and used to make decisions about, some portion of the earth(DeMers, 1997).

⑨A working GIS integrates five key components: hardware, software, data, people, and methods (ESRI, 1997).

⑩A computer system for analyzing and mapping just about anything, moving or stationary (Lang, 2000).

⑪Much more than a container of maps in digital form. A GIS is also a computerized tool for solving geographic problems, a spatial decision support system; it is a useful tool for revealing what is otherwise invisible in geographic information (Longley, 2005).

1.3.3 Alternative Names of GIS

Alternative names people have used over the years illustrate the range of applications and emphasis, such as: A GIS -automated geographic information system, AM/FM-automated mapping and facilities management, CAD-computer-assisted drafting, CAM-computer-assisted mapping (or manufacturing), environmental information system, image-based information system, knowledge-based GIS, LIS-land information system, multipurpose *cadaster* (地籍图), multipurpose geographical data system, planning information system, spatial data handling system, spatial information system, and so on.

1.4 Contributing Disciplines and Technologies

GIS is a *convergence* (汇聚) of technological fields and traditional disciplines. It has been described as an *enabling technology* (使能技术) because of the potential it offers for various fields which need the support of spatial data. Each related field provides some unique techniques for GIS. Moreover, many of these related fields give priority to data collection-GIS brings them together by emphasizing integration, modeling and analysis. As the integrated field, GIS is often viewed as the science of spatial information (作为集成化领域, GIS常被认为是空间信息的科学).

①Geography is generally concerned with understanding the world and it provides techniques for conducting spatial analysis and a spatial perspective on inquiry.

②*Cartography* (地图学) is concerned with displaying spatial information. It provides *extensive traditions* (大量惯例) in the design of the maps which are currently the main source of input data of GIS and an important form of GIS output. Computer cartography (also called digital cartography or automated cartography) provides methods for digital representation of cartographic features and methods of visualization of the real world.

③Remote sensing images from space and the air (*Aerial photography and satellite images*, 航空影像和卫星图像) are an important source of geographic data. Remote sensing includes techniques for data acquisition and processing anywhere on the globe at low cost, consistent update potential. Numerous image analysis systems contain sophisticated analytical functions; interpreted data from a remote sensing system can be merged with other data layers in a GIS.

④*Photogrammetry* (摄影测量学) is *at the root of* (根源) most data on topography (ground surface elevations) used for input to GIS. It is the science and technology for making precise spatial measurements from aerial photos.

⑤*Geodesy* (大地测量学) is a source of highly accurate positional control for GIS; Surveying offers a wide range of high-quality topographical data on locations.

⑥Statistics is important in understanding issues of error and uncertainty in GIS data. Many models built by using GIS are statistical in nature; many statistical techniques are employed for analysis.

⑦Computer Science: *computer-aided design* (CAD) provides software and techniques for data input, display, visualization and representation, particularly in 3 dimensions. Advances in *computer graphics* provide hardware, software for handling and displaying graphic objects, techniques of visualization. *Database management systems* (DBMS) contribute methods for representing data in digital form, procedures for system design and handling large volumes of data, particularly access and update. *Artificial intelligence* (AI, 人工智能) resorts to the computer for choice-making based on available data in a way seen to emulate human intelligence—the computer can act as an "expert" in such functions as designing maps, generalizing map features. GIS has yet to take full advantage of AI, but AI has already provided methods and techniques for system design. *Virtual reality* (VR, 虚拟现实) is a computer-simulated environment that can simulate the physical presence in places in the real world or imagined worlds. Since GIS technology has played an instrumental part in the real-time simulation process, the use of VR technology would only be logical to exploit GIS technology.

⑧Several branches of mathematics, especially geometry and graph theory, are employed in the GIS system design and analysis of spatial data.

1.5 Brief History of GIS

1.5.1 Era of Open Innovation

The first applications of spatial analysis dates back to 1832. The French geographer *Charles Picquet* represented the 48 districts of the city of Paris by *halftone color gradient* (半彩色梯度) according to the percentage of deaths by *cholera* (霍乱) per 1000 inhabitants. Shortly afterward, John Snow depicted a cholera outbreak in London in 1854 using points to represent the locations of some individual cases and analyzed the clusters of geographically dependent phenomena in a map, which is possibly the earliest use of a geographic methodology in *epidemiology* (流行病学).

With the development of *photozicograhpy* (照相锌版术) in the early 20th century, maps could be allowed to be split into different layers, which means that a separate layer could be worked on without the other layers to confuse the *draughtsman* (制图员). This work was originally drawn on *glass plates* (玻璃板), but the later *plastic film* (塑胶膜) was introduced. When all the layers were finished, they were combined into one image using a larger process camera. Once color printing came in, the layers idea was also used for creating separate printing plates for each color. While the use of layers much later became one of the main typical features of a contemporary GIS, the photographic process just described is not considered to be a GIS in itself-as the maps were just images with no database to link them to.

With the advent of computer, general-purpose computer mapping applications appeared in the early 1960s. The world's first true operational GIS-CGIS (Canada Geographic Information System) was developed as a mainframe-based system by Dr. *Roger Tomlinson* in the mid-1960s to assist in regulatory procedures of land-use management and *resource monitoring* (资源管理). CGIS represented an improvement over "computer mapping" applications as it provided capabilities for overlay, measurement, and digitizing/scanning. It supported a national coordinate system that spanned the continent, coded lines as arcs having a true embedded topology. It stored attributes and locational information in separate files. Therefore, *Roger Tomlinson* was known as the Father of GIS. CGIS lasted until the 1990s and built a huge digital land resource database in Canada. However, the CGIS was never available commercially.

In parallel, *Howard T. Fisher* founded the Laboratory for Computer Graphics and Spatial Analysis at the Harvard Graduate School of Design in 1964, where a number of important theoretical concepts in spatial data handling were developed. By the 1970s it had distributed profound software code and systems, such as SYMAP, GRID, and ODYSSEY, that served as sources for subsequent commercial development — to universities, research centers and corporations worldwide.

1.5.2 Era of Commercialization

The era of commercialization was roughly from 1980s to 1999s. From the early 1980s on, numerous *commercial vendors* (商业销售商) of GIS software emerged, including M&S Computing (later Intergraph) along with Bentley Systems Incorporated for the CAD platform, ESRI (Environmental Systems Research Institute), CARIS (Computer Aided Resource Information System), and ERDAS (Earth Resource Data Analysis System). Their GIS software products successfully incorporated many of the CGIS features, combined the first generation approach to separation of spatial and attribute information with a second generation approach to organizing attribute data into database structures. In parallel, the development of two public domain systems (MOSS and GRASS GIS) began in the late 1970s and early 1980s.

In 1986, the leading desktop GIS product MIDAS (Mapping Display and Analysis System) emerged for the DOS operating system. This was renamed to MapInfo for Windows in 1990 when it was ported to the Microsoft Windows platform. This began the process of shifting GIS from the research department into the business environment.

With the advent and popularity of powerful PCs, the market of GIS software continued to grow, and the GIS software industry was also showing *unprecedented* (空前的) development trend.

By the end of the 20th century, the rapid growth in various systems had been *consolidated* (稳定) and standardized on relatively few platforms. Users were beginning to explore viewing GIS data over the Internet, requiring data format and transfer standards. More recently, a growing number of free, open-source GIS packages run on a range of operating systems and can be *customized* (定制) to perform specific tasks. Increasingly geospatial data and mapping applications are being made available via the *World Wide Web* (万维网).

1.5.3 Era of Exploitation

With more and more GIS applications explored, GIS (or GIS-related) technologies have been revolutionized. Modern history of GIS exploitation has since ushered in a new era.

With the development of network technology, the computer network is becoming more and more powerful, and it provides the public a *novel* (新奇的) tool to recognize and understand the world better. And then, application of GIS technology in the network is getting more and more popular worldwide. This ever-changing technology gradually recommends itself to the public (这一发展日新月异的技术逐渐被大众所接受). Each *breakthrough* (突破性进展) in the field of GIS technology encourages the growth of related industries. Simultaneously, GIS is developing along with technologies of relevant operational areas. Today, the GIS and the network have been heavily *intertwined* (缠绕).

In general, the history of GIS is much more complex than described in this brief history. The selected major events of the past three eras are summarized in the Table 1-3.

Table 1-3 The Brief History of GIS

Dates	Events and Notes
The Era of Innovation	
1957	Swedish meteorologists and British biologists worked out the earliest known automated mapping
1963	*Roger Tomlinson* led the first GIS project-CGIS-completed in 1971. This project pioneered much technology and introduceed the term GIS
1963	Founded in the US, the URISA (Urban and Regional Information Systems Association) soon became the communication platform for GIS innovators
1964	*Howard Fisher* founded the Laboratory for Computer Graphics and Spatial Analysis at Harvard University and developed the original software for automatic cartography-SYMAP (maps were generated with textual characters-text mode) in 1966
1967	US Bureau of Census developed DIME-GDF (Dual Independent Map Encoding-Geographic Database Files), a data structure and street-address database for 1970 census

(续)

Dates	Events and Notes
	The Era of Innovation
1967	UK ECU (Experimental Cartography Unit) was founded and then pioneered in a range of computer cartography and GIS areas
1969	*Jack Dangermond*, a student from the Harvard Lab, and his wife *Laura* founded the Environmental Systems Research Institute-ESRI Inc (developers of ArcGIS software)
1969	*Jim Meadlock* and four others that worked on guidance systems for Saturn rockets projected M&S Computing, later renamed Intergraph Corp (developers of Geomedia software)
1969	Edited by *Ian McHarg*, *Design With Nature* was published. It was the first to describe many of the concepts in modern GIS analysis, including the map overlay process
1969	Edited by *Nordbeck* and *Rystedtf*, the first technical GIS textbook detailed algorithms and software, which were developed for spatial analysis
1972	Originally named ERTS (Earth Resources Technology Satellite), Landsat 1 was the first of many major earth remote sensing satellites launched
1973	First digitized production line was set up by Ordnance Survey, Britain's national mapping agency
1974	AutoCarto 1 Conference was held in Reston, Virginia. It was the first important of the series of conferences that set the GIS research agenda
1976	GIMMS was written by *Toni Waugh* (a Scottish academic). This vector-based mapping and analysis system is in worldwide use now
1977	Harvard Lab organized a major conference-Topological Data Structures conference and developed the ODYSSEY GIS
1977	MOSS (Map Overlay and Statistical System), a GIS software technology, came out in the late 1977 and was first deployed for use in 1979. MOSS represents a very early public domain and open source GIS development, predating the better known GRASS by 5 years
1979	*Siemens Nixdorf* developed the GIS software SiCAD
1979	CARIS (Computer Aided Resource Information System) company was founded in Canada to develop and support geomatics software for land and marine applications
	The Era of Commercialization
1981	ArcInfo, the first major commercial GIS software system, was launched. Designed for minicomputers and based on the vector and relational database data model, it established a new standard for the industry
1982	Autodesk Company was founded (AutoCAD developers)
1984	The first accessible source of information about GIS *Basic Readings in Geographic Information Systems*, by *Duane Marble*, *Hugh Calkins*, and *Donna Peuquet*, was published
1984	ESRI released the revolutionary concept of linking databases records as attributes to each map element
1984	Bentley Systems (division of Intergraph) launched software platform MicroStation for CAD & GIS
1985	The GPS (Global Positioning System) gradually became a major source of data for navigation, surveying, and mapping
1986	Edited by Peter Burrough, *Principles of Geographical Information Systems for Land Resources Assessment* was published. It was the first specifically on GIS principles and quickly became a worldwide reference text for GIS students

(续)

Dates	Events and Notes
The Era of Commercialization	
1986	Founded by 4 students, *Laszlo Bardos*, *Andrew Dressel*, *John Haller*, and *Sean O'Sullivan* at nearby Rensselaer Polytechnic Institute, MapInfo Corp. developed the first desktop GIS. It defined a new standard for GIS products, complementing earlier software systems
1986	The U. S. Army Research Laboratories launched the development of GRASS (Geographic Resources Analysis Support System)-one of the most appreciated GIS software nowadays
1987	*Terry Coppock* and others published the first journal on GIS-*International Journal of Geographical Information Systems* (now *IJGI Science*). The first journal collected papers from the USA, Canada, Germany, and UK
1987	Chorley Report—*Handling Geographical Information* was an influential report from the UK government that highlighted the value of GIS
1988	Smallworld Company was founded (nowadays-GE Smallworld-Smallworld software developers)
1988	GISWorld, now GeoWorld, the first worldwide magazine dedicated to GIS, was published in the USA
1988	TIGER (Topologically Integrated Geographic Encoding and Referencing), a follow-on to DIME, was described by the US Census Bureau. Low-cost TIGER data stimulated rapid growth in US business GIS
1988	Two separate initiatives, the US NCGIA (National Center for Geographic Information and Analysis) and the UK RRL (Regional Research Laboratory) Initiative showed the rapidly growing interest in GIS in academia
1990	The French GeoConcept Company was founded (GeoConcept developers)
1991	*Big Book* 1 was published. Substantial two-volume compendium Geographical Information Systems: principles and applications, edited by David Maguire, Mike Goodchild, and David Rhind documents progress to date
1992	Sponsored by the US Defense Mapping Agency (now NGA), 1.7 GB DCW (Digital Chart of the World) was released. It was the first integrated 1 : 1 million scale database offering global coverage
1994	Signed by President *Clinton*, Executive Order 12 906 led to the creation of US NSDI (National Spatial Data Infrastructure), clearing houses, and FGDC (Federal Geographic Data Committee)
1994	OpenGIS© Consortium (Open Geospatial Consortium) consisting of more than 220 GIS companies was established to improve interoperability, standards for geospatial content and services, GIS data processing and data sharing
1995	The first complete national mapping coverage: Great Britain's Ordnance Survey completed the creation of its initial database-all 230 000 maps covering country at the largest scale (1 : 1250, 1 : 2500 and 1 : 10 000) encoded
1996	Several companies, notably Autodesk, ESRI, Intergraph, and Mapinfo, released a new generation of Internet-based GIS products at about the same time
1996	MapQuest-Internet mapping service was launched and produced over 130 million maps in 1999. Subsequently it was purchased by AOL for $1.1 billion
1999	The first GIS Day attracted over 1.2 million global participants who shared an interest in GIS
The Era of Exploitation	
1999	Launch of a new generation of satellite sensors: IKONOS claimed 90 centimeter ground resolution; Quickbird (launched 2001) claimed 62 cm resolution
2000	Industry analyst *Daratech* repored GIS hardware, software, and services industry at $6.9 bn, growing at more than 10% per annum
2000	GIS had more than 1 million core users and approximately 5 million casual users

(续)

Dates	Events and Notes
	The Era of Exploitation
2002	Online US national-scale atlas was launched (www. nationalatlas. gov)
2003	Online national statistics for the UK which describe economy, population, and society at local and regional scales was launched (www. statistics. gov. uk)
2003	A US Federal E-government initiative provided access to geospatial data and information (www. geodata. gov/gos)
2004	The biggest GIS user in the world, National Imagery and Mapping Agency (NIMA), renamed NGA to signify emphasis on geo-intelligence
2004	Google Earth, originally called EarthViewer 3D, was created by Keyhole, Inc. The CIA-funded (Central Intelligence Agency) company was taken over by Google. The product was re-released as Google Earth in 2005
2005	Provided by Google, Google Maps (formerly Google Local), embedded in third-party websites via the Google MapsAPI, provided many map-based services, and maps
2005	A new and authoritative geodetic height determination of Mount Qomolangma (8844. 43±0. 21 m) was conducted by the State Bureau of Surveying and Mapping (SBSM) of China
2006	OpenLayers, the completely free Open Source JavaScript, was released by MetaCarta
2007	INSPIRE (Infrastructure for Spatial Information in Europe) directive entered into force
2008	GeoEye-1 was launched. It provided Google with color geospatial imagery at half-meter ground resolution
2008	Google Earth for the iPhone was released; developed initially for Google Earth to exchange geographic information and mapping presentations (Google Earth KML), the file format is now an international standard
2009	Bing Maps (previously Live Search Maps, Windows Live Maps, Windows Live Local, and MSN Virtual Earth) from Microsoft was released
2009	"Smarter Planet" campaign was initiated by IBM
2009	Toyota Motor Sales announced a new in-car convenience telematics program for select
2010	ArcGIS 10 was released by ESRI. It provided new ways to share information, supplied GIS in the Cloud
2011	Operated by the Russian Aerospace Defence Forces, GLONASS (Global Navigation Satellite System) achieved global coverage with a fully orbital constellation of 24 satellites
2012	Ziyuan 3 (ZY-3), the Chinese Earth observation satellite was launched
2013	GeoEye Inc. (formerly Orbital Imaging Corporation or ORBIMAGE), an American commercial satellite imagery company, was merged into the DigitalGlobe corporation
2013	GitHub announced to automatically view any GeoJSON files that may be in a repository inside an interactive map driven by MapBox technology
2015	The first research report on the development of geographic information industry in China was announced
2017	The first meeting of *the Steering Committee of the United Nations World Conference on Geographic Information* (联合国世界地理信息大会指导委员会) was held in Wuhan of China
2018	The General Plan of the 3rd National Land Survey in China was officially released
2018	The Chinese National Geodetic Coordinate System 2000 was in full use, and the basic surveying and mapping results using 1954 Beijing coordinate system and 1980 Xi'an coordinate system came to a complete stop in 2019
2020	BeiDou Navigation Satellite System of China was firstly officially opened

Note: Partially according to Longley et al., 2005.

1.6 Component of a GIS

The fundamental six parts of an operational GIS are: hardware, software, people, data (or database), network and procedures(Figure 1-1).

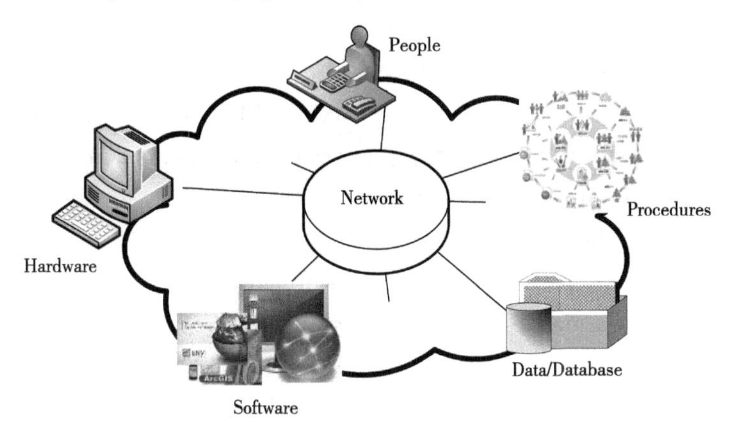

Figure 1-1 Components of a GIS

1.6.1 Hardware

The basic GIS *ingredient* (组分, 要素) is the *hardware* for storing, processing, transmittingg, and displaying geographic information or spatial data. Traditionally, GIS hardware environment specifically refers to a desktop computer coupled with some external devices including: input devices-digitizers, scanners and *measuring instruments* (测量设备); output devices-plotters, printers and high-resolution display devices; data storage and transfer devices-tape drives, CD-ROM drive, *portable hard drives* (移动硬盘) and *hard disk arrays* (硬盘阵列); and network devices including *wiring systems*, *bridges*, *routers* and *switches* (布线系统、网桥、路由器和交换机). Nowadays, GIS functions can be implemented through more convenient hand-*type* (便携式的) devices, such as laptops, *personal data assistants* (PDAs), *in-vehicle devices* (车载移动电话), and even *cellular telephones* (携带式移动电话).

1.6.2 Software

GIS software is the core of the GISystem used to perform various GIS functional operations, including data input, processing, database management, spatial analysis and graphical user interface (GUI). Locally, it runs in the user's machine. The GIS software *installation system* (安装系统) is illustrated by Figure 1-2.

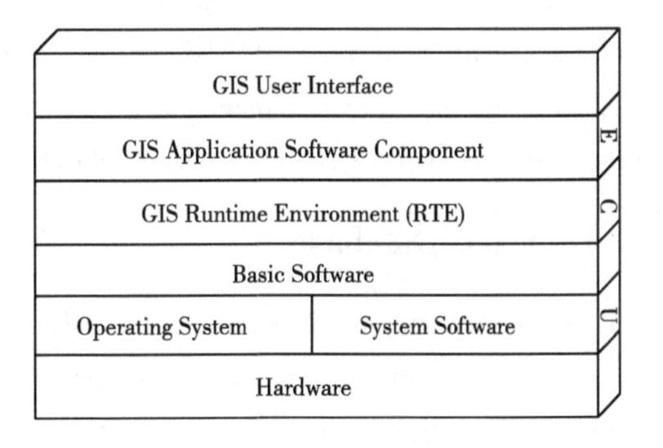

Figure 1-2 GIS Software Installation Framework

An *operating system* (OS) is a set of software that manages computer hardware resources and provides common services for computer programs. It is a vital component of the system software. The GIS professional software also requires an operating system to work. The popular operating systems include Microsoft Windows, MS-DOS, Unix, Linux, Mac OS, Android, iOS, Windows Phone, and IBM z/OS, etc.

The system software is designed to *operate and control* (操控) the computer hardware. For instance, device drivers such as computer BIOS and device *firmware* (固件). The operating system allows the parts of a computer to *work jointly* (联合开展工作) by performing tasks like transferring data between memory and *disks* (磁盘) or rendering output onto a display device. It also serves as a platform to run high-level system software and application software.

The basic software is the standardized software layer, which provides services to the GIS software components and is necessary to run the functional part of the software. It does not fulfill any functional job itself and is situated below the GIS *runtime environment* (RTE, 运行环境). The basic software contains standardized and *engine control unit* (ECU, 引擎控制单元) specific components.

At the system design level, the GIS runtime environment acts as a communication center for inter-and intra-ECU information exchange. As the communication requirements of the software components running in the RTE are application dependent, the RTE needs to be *tailored* (特制的), partly by ECU-specific generation and partly by the configuration. Thus, the resulting RTE will differ between one ECU and another.

GIS application software components consist of GIS professional software and database software that are mapped on the ECU and routed through the GIS runtime environment. GIS user interface assures the connectivity of GIS application software elements surrounding the GIS runtime environment. There are numerous different GIS software packages available today. All packages must be capable of data input and *verification* (确认), storage and database management, transformation and manipulation, query and analysis, output and presentation, but the appearances, methods, resources, and ease of use of the various systems may be very different. More likely it is a package bought from one of the GIS vendors. Each vendor offers a range of products, designed for different levels of *sophistication* (精密, 复杂), different volumes of data, and different application *niches* (用户群, 环境). The GIS database software specifies the database to support complex spatial data management and non-spatial attribute services, such as: MySQL, Microsoft SQL Server, Oracle, dBASE, FoxPro, IBM DB2, Sybase, InforMIX and so on.

1.6.3 Data/Database

It is the most important component of a GIS. GIS will integrate spatial data with other data resources and use database management system (DBMS) to organize and maintain the spatial data. Perhaps the most time-consuming and costly aspect of initiating a GIS is to create and maintain a database, which consists of a digital representation of selected aspects of some specific area of the

Earth's surface or near-surface (Figure 1-3). In a GIS, the data is stored in a database in the form of structured data, which can be purchased from a commercial data provider or collected or compiled according to *custom specification* (用户自定义的规范).

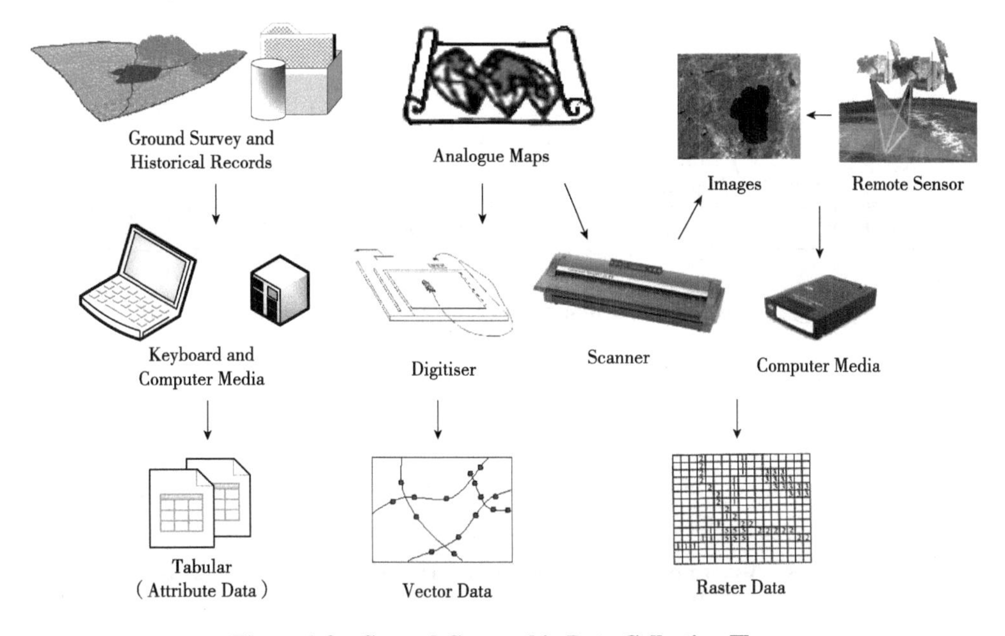

Figure 1-3 General Geographic Data Collection Flow

1.6.4 Network

With the development of computer networks and distributed computing technology, demand for information exchange in a wide variety of applications has been increasing in recent years. As a component of GIS, GIS computing resources distributed over a computer network are connected and integrated to allow rapid communication, digital information sharing and the better deployment of geographic data. Doubtlessly, GIS relies increasingly on the network for information exchange, and the network has proven very popular for delivering GIS applications.

More detailed info. please check the course of WebGIS.

1.6.5 Procedures

In addition to these four components, a GIS also requires management. An organization must establish fitting procedures. In a sense, the *rules and regulations must be adhered to in handling affairs* (无规矩不成方圆) to ensure that the GIS activities *stay within budgets* (控制预算), maintain high quality, and generally meet the needs of the organization.

As the key to successful operation of GIS technology(程序是 GIS 技术成功应用的一个关键), procedures involve how the data will be retrieved, input, stored, managed, transformed, analyzed, and finally presented in a final output(Figure 1-4).

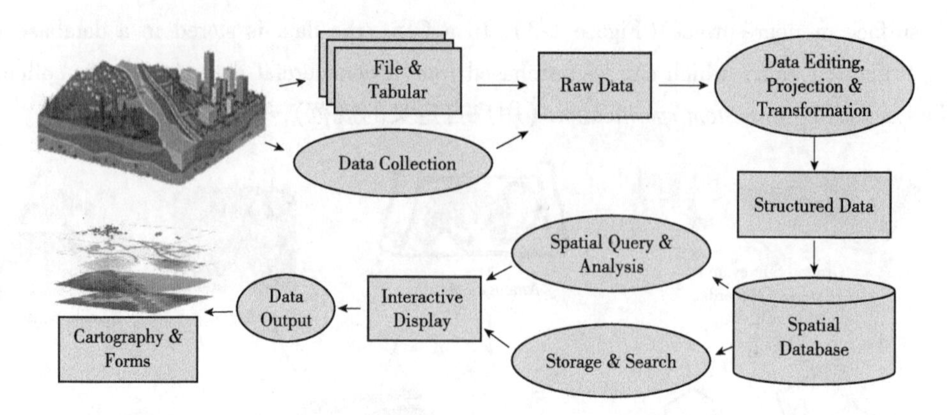

Figure 1-4 Common Procedures of a GIS Project

1.6.6 Personnel

As the most active and energetic element actually making the GIS work, the personnel determines the operation mode of GIS and the expression of information (人是最活跃的、使 GIS 真正运行的要素，决定着 GIS 系统的工作方式和信息的表达方式). GIS personnel includes a wide range of positions. According to the roles in a GIS, GIS personnel can be broadly divided into three categories:

GIS end users (最终用户): theymay not be the GIS specialists, or needn't to know *the technical content or the operating mechanism* (技术内涵或运行机制). What they need is simply a simple operation of GIS. They may utilize GIS to *browse referential material* (浏览参考资料), conduct business, perform professional services, or make decisions. The majority of GIS personnel are end users: ordinary users, managers, planners, engineers, lawyers, doctors, *business entrepreneurs* (企业家), etc.

GIS professionals/specialists: they have received formal training in GIS with some basic GIS theory and practical experience. Engaged in GIS technology research, they are GIS (scientific) researchers, GIS educationists, GIS managers, database administrators, application specialists, systems analysts, programmers and so on.

GIS Developers/System Integrators: they refer in particular to GIS organization and enterprise devoted to software development, *system integration* (系统集成), such as ESRI Inc., Intergraph Inc., Autodesk Inc., SuperMap Co. Ltd., etc.

1.7 GIS = GISystem, GIScience, GIService and GIStudies

Geo-Information research and technology has experienced developments over five decades. Every important GIS development was driven by a significant breakthrough of the mainstream information technology. Thus, GIS-related meanings have evolved from a computer-based system to an internet-or web-based service, from a technologically dominated view to an increasingly science-oriented view, from a professional-oriented application to the public-oriented application.

GISystem is designed to perform certain specific functions, which allow users to analyze, edit spatial data and present the results.

GIScience focuses on the scientific principles that GISystems are based upon, dealing with fundamental organizing principles, data models, *algorithms* (算法), and methods for representing geography and spatial relationships with the use of GISystems and related technologies. In a sense, GISystems has given birth to GIScience (地理信息系统孕育了地理信息科学) and GIScience drives and guides the development of GISystems (地理信息科学驱动和引领着地理信息系统的发展).

With the development and popularity of the network technology, the Internet and the Web have allowed GIS users to access specific functions e. g. Google Maps and Google Earth. The *marriage* (紧密结合) of network and GISystems has brought about a "service-oriented" approach to GIS rather than "system-oriented", namely GIServices. Instead of owning a GISystem, GIS people can be served by GIS functionalities from remotely distributed GIServices resources.

More detailed info. can be checked at the course of "WebGIS".

GIStudies concerns studying the impacts of geographic information and technologies on society, how the geographic information and technology are integrated into a societal context.

1.8 Benefits of GIS

GIS benefits organizations of all sizes and in almost every industry. There is a growing awareness of the economic and strategic value of GIS. According to ESRI, the benefits of GIS generally can fall into five basic categories:

①Cost Effectiveness and Increased Efficiency (节约成本和提高效率): GIS is used extensively to optimize maintenance schedules and daily fleet movements. Typical implementations can result in a saving of 10 to 30 percent in operational expenses through reduction in fuel use and staff time, improved customer service, and more efficient scheduling.

②Better Decision Making (更好地决策): GIS is the preferred technology for making better decisions about location. Common examples include *real estate site selection* (房地产选址), route/corridor selection, evacuation planning (疏散计划), conservation, *natural resource extraction*(自然资源开采), etc. Making correct decisions about location is critical to the success of an organization.

③Improved Communication (加强沟通): GIS-based maps and visualizations *greatly assist in*(帮助) understanding situations and in story telling. They are a type of language that improves communication between different teams, departments, disciplines, professional fields, organizations, and the public.

④Better Recordkeeping (更好地保存记录): Many organizations have a primary responsibility of maintaining *authoritative*(权威性的) records about the status and change of geography.

GIS provides a strong framework for managing these types of records with full *transaction*(交易；事务) support and reporting tools.

⑤Managing Geographically (地理化管理)：GIS is becoming essential to understand what is happening and what will happen in geographic space. Once we understand, we can *prescribe*(指示，规定) actions. This new approach to management-managing geographically-is transforming the way that organizations operate.

With the increasing importance and wide use, the GIS is becoming an essential tool for geographical study and practices. At the same time, Geographical data is becoming more widely available and less expensive so that using such data efficiently and wisely gives *competitive margin*(竞争优势) for a business. The GIS will eventually make your study in geography easier, more efficient and enjoyable. After all, the GIS is already in your life, whether you like it or not!

Vocabulary

alternative[ɔːlˈtɜːrnətɪv] *adj.* 两者择一的；供选择的；非主流的 *n.* 二者择一；供替代的选择
ambiguous[æmˈbɪɡjuəs] *adj.* 模棱两可的；含糊不清的
arena[əˈriːnə] *n.* 竞技场
association[əˌsoʊʃiˈeɪʃn] *n.* 协会；社团；交往；联想；联合；结合
atmosphere[ˈætməsfɪə(r)] *n.* 大气圈；空气；气氛；气压
biosphere[ˈbaɪəʊsfɪə(r)] *n.* 生物圈
cartography[kaːrˈtaːɡrəfi] *n.* 地图制作；制图学；制图法
compendium[kəmˈpendiəm] *n.* 简要；概略；提纲；手册
consortium[kənˈsɔːʃɪəm] *n.* 财团；联合；合伙
continent[ˈkɒntɪnənt] *n.* 大陆；洲；(the Continent)欧洲大陆
convergence[kənˈvɜːdʒəns] *n.* 收敛；汇聚；汇合点
disciplines[ˈdɪsəplɪn] *n.* 纪律；训练；学科 *vt.* 训练；惩罚
GUI graphical user interface 的缩写，图形用户界面
geodesy[dʒiːˈɔdəsɪ] *n.* 测地学；大地测量学
hydrosphere[ˈhaɪdroʊsfɪr] *n.* 水圈；水界；水气
intertwine[ˌɪntərˈtwaɪn] *v.* 纠缠；缠绕；编结
lithosphere[ˈlɪθəsfɪr] *n.* 岩石圈；地壳
pedosphere[ˈpedəsfɪə] *n.* (地球的)土壤圈；土界
perspective[pərˈspektɪv] *n.* 远景；看法；透视 *adj.* 透视的
philosophy[fəˈlaːsəfi] *n.* 哲学；哲理
retrieve[rɪˈtriːv] *vt.* 恢复；挽回；取回 *vi.* 找回猎物 *n.* 恢复；取回；救出险球
seamless[ˈsiːmləs] *adj.* 持续的；无缝的
segment[ˈseɡmənt] *n.* 部分；弓形；瓣；段；节 *vt.* 分割
sophisticated[səˈfɪstɪkeɪtɪd] *adj.* 老练的；精密的；复杂的；久经世故的

topology[tə'pɔlədʒi] *n.* 拓扑学；地志学；布局解剖学
valiant['væliənt] *adj.* 勇敢的；英勇的 *n.* 勇士；勇敢的人

Questions for Further Study

1. How to understand the relationship between geomantic omen (堪舆) and geography?

2. Examine the geographic data available for the province, county, city, town or rural area where you live or study. Use it to produce a short illustrated profile of either the socioeconomic or the physical environment.

3. What is a GIS? Try your best to find out alternative names of GIS and why are there so many names for GIS?

4. What does a GIS consist of? What can we answer with the help of GIS?

5. What is the relationship between the traditional map and GIS?

6. Try your best to figure out the current status of GIS publications, GIS educations and GIS industries.

Chapter 2

GIS Software

GIS software is a very important component of an operational GIS. It provides storage, display, and analysis of geographic data. It is a data processing engine composed of a complete set of computer programs with geographic processing functions (GIS 软件可对地理数据进行存储、显示和分析，它是由一整套具有地理处理功能的计算机程序而组成的数据处理引擎).

2.1 Evolution of GIS Software

By surveying the whole history of GIS software development, the evolution of GIS software can be roughly classified into 4 generations.

2.1.1 The 1st Generation of GIS

From the middle 1960s to the late 1980s, the first generation of GIS software came into being. GIS software in the formative years contained a few collections of computer routines that only a skilled programmer could use to build a working GIS. The basic technical characteristics include: ①The standard means of communicating with a GIS was to type instructions in command lines; ②Layer was the basis for handling. Related operations, such as overlay, buffering, path analysis, etc. were limited to the current layer; ③System was taken as the center-each GIS software was unique in terms of its capabilities and different GIS software systems was largely unrelated or independent; ④GIS software was installed by *single users for single processing* (单机单用户); and ⑤Data (spatial data and attribute data) management wass typically implemented by the use of the file system directly.

2.1.2 The 2nd Generation of GIS

The period from the late 1980s to the middle 1990s witnessed the high speed development and promotion of GIS software applicationwith the development of software engineering techniques and the growth of GIS market. As a software tool, GIS-related theories and technologies gradually came to maturity. At the same time, the rapid development of Internet/Intranet provided various demands on the GIS software. However, no fundamental change took place in the basic technical structure of GIS software: ①Layer was still the basis for an operational GIS; ②Data (attribute data in particular) could be transferred among different systems through interchange data format; ③Multi-user online processing based on the Internet/Intranet emerged; ④Redevelopment ability was strengthened; ⑤Attribute data could be managed by commercial database management system, but spatial data management was still mainly implemented by the file system; ⑥GIS software application domains have been widely extended although it was still dominated by the function of data management.

In this period, two key developments occurred to make the GIS software easier to use and more generic. ①*Command line interfaces* (命令行界面) were supplemented and eventually largely replaced by *graphical user interfaces* (GUIs, 图形用户界面). These menu-driven, form-based interfaces greatly simplified user interaction with a GIS. ②Some custom features were developed by applying a *high level programming language* (高级编程语言) to *application programming interfaces* (APIs, 应用程序接口) for specific-purpose applications from the generic toolboxes. In particular, the feature of creating custom application solutions allowed developers to initiate wider applications for end users in specific markets or areas, such as resources management, governments, utilities, military, and environment. Simultaneously, additional terms were developed to distinguish these subtypes of GIS software: planning information systems, automated mapping/facility management (AM/FM) systems, land information systems, location-based services systems, and so on.

2.1.3 The 3rd Generation of GIS

The period from the middle 1990s to the late 2000s witnessed the extremely speedy expansion of GIS application. ①The openness of GIS software system was significantly increased; and component-based technology had gradually been the mainframe during the development of GIS software; ②The spatial data was still mainly the static and 2-dimensional data, and it was *mono temporal data* (单时相数据); ③Technologies concerning metadata, data sharing, geographic services sharing and system interconnect were widely developed; and ④Great importance was continuously attached to the standardization of GIS functionality and interoperability (GIS 功能标准化和互操作得到不断重视).

2.1.4 The 4th Generation of GIS

Since the late 2000s, GIS software has gradually obtained: ①Powerful integration capabilities with different systems such as RS, GPS, MIS (Management Information System), OA (Office Automation), CAD, DCS (Distributed Control System), etc.; ②Certain capacity of processing 3-demension data and time series data; ③*Multimodal user interfaces* (多通道用户界面); ④Functionality of virtual reality presentation, distributed spatial data management and computing.

2.2 Architecture of GIS Software

2.2.1 Scope of GIS Projects

Originally, most organizations carried out *single-purpose projects*（单一项目）with the help of a GIS, although each project involves different hardware, software, data, people, and procedures. Characteristics of this modest scope are: ①GIS data and experience sharing is difficult; ②The logical and expected result is only a project-specific output; ③The project is almost of one-time effort and the project has an end date; ④The project needs to pay the acquisition cost and no long-term support as expected; and ⑤Single database established for the project is primarily used by one user.

To save money and encouragesharing, several projects in the same department may be carried out simultaneously. In particular, a centralized database formed and was used by multiple users. The second level of GIS scope — a *departmental-level*（院系级）GIS application came into being. This often led to the creation of common standards, development of a specialized GIS team, and *procurement*（获得）of new GIS capabilities. Yet it was also quite common for different departments to have their different GIS software and *data standards*（数据标准）respectively.

As GIS became more *pervasive*（普遍的）, a great number of users in an organization have realized that GIS is a useful way to structure many of the organization's assets, processes and workflows. Finally, with a long-term vision and efforts, an *enterprise-level*（企业级）GIS has been integrated in an organization, and distributed database is shared on the network so that users can manage, share, and use spatial data and related information to address a variety of needs.

2.2.2 Three-tier Architecture

The three-tier model is a software architecture pattern in whichthe user interface, the tools, and the data management system are developed and maintained as independent modules (Figure 2-1).

The data is the basis of GIS and the *data access*（数据访问）is the *premise*（前提）of the GIS software for data management and analysis. The data (information) is stored and retrieved from file system or a database management system (DBMS).

The GIS toolset exists in the *business logic layer*（业务逻辑层）, which consists of one or more separate modules available for moving and processing geographic data to *accomplish*（实现）a GIS task. Commonly, these tools are classed according to their functions, shown as a *graphical button*（图形按钮）.

The user interface (or GUI) is a *presentation layer*（表现层）of GIS software, across which users can exchange information via the keyboard, mouse, menus of a computer system or other *controls*（控件）. Normally, the interface can be broken down into four basic features: map, layers, navigation and functions (Figure 2-2). The map is much like a paper map but displays on the computer screen. Also, maps can be integrated with a GIS. Layers often represent different

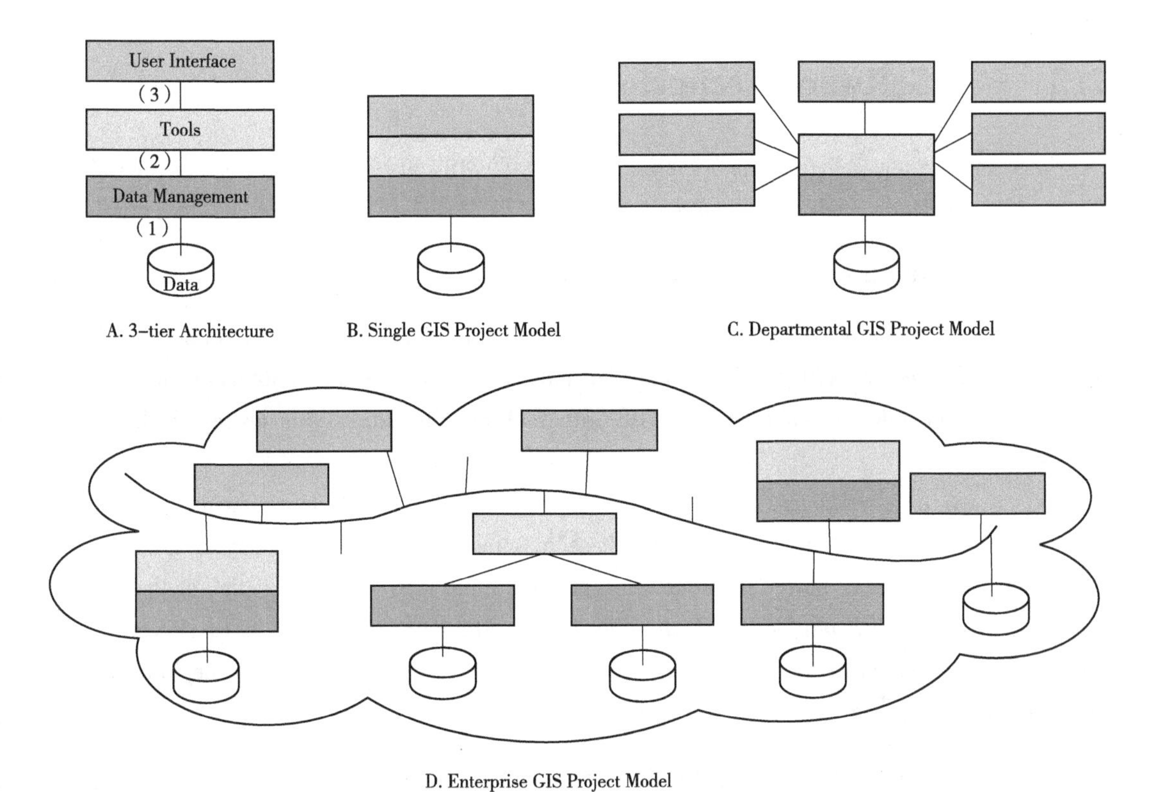

Figure 2-1 Three-tier Architecture with 3 Different GIS Project Models

types of maps, for example topographical maps, thematic maps or point maps. Unlike a paper map, the map in a GIS is not static, it can be navigated and zoomed in or out. The functions are the tools which allow us to do different things with the data in our maps, from making layers more transparent, measuring distances and areas, viewing the data three-dimensionally to complex statistics.

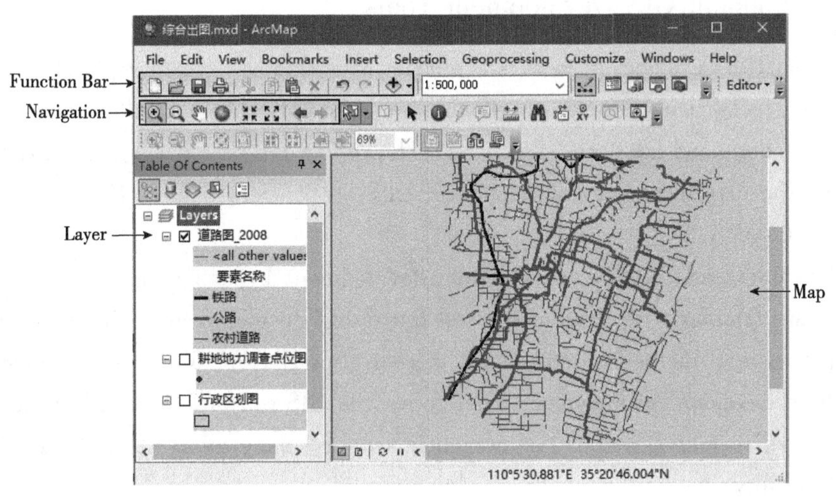

Figure 2-2 GUI Example of ArcMap Software

2.3 GIS Software Technology

GIS software technology is the organizational form of GIS software, which *bears on* (决定着) the overall *application mode and integrated efficiency* (应用方式和集成效率) of GIS software.

2.3.1 Integrated GIS Development Tools

In the early stage of GIS software, a lot of GIS functional modules came into being gradually to meet specific needs. With the increasing demand, researchers began to integrate these disparate functional modules as an integrated GIS with multiple functions. This process is known as integration.

Integrated GIS development tools (集成式 GIS 开发工具) is a GIS development package with a collection of various functional modules. These functional modules was taken as a completely independent system with powerful capabilities of data input and output, spatial analysis, and *good features of graphics platform and reliability* (良好的图形平台和可靠性能). However, the GIS software developed in this way was complex, large, costly, and difficult to be integrated to other application systems.

2.3.2 Modular GIS Development Tools

In order to overcome the shortcomings of integrated GIS, researchers have tried to develop the next generation of GIS software based on the overall system architecture and the *functional correlations* (功能的关联度). That is, GIS software runs in accordance with different function modules. Therefore, the GIS software developed in this way is highly flexible and easy to *develop and expand for the second time* (二次开发).

2.3.3 Component GIS Development Tools

Component GIS development tools is a product of computer technology development, and it represents the development direction of GIS software. Not only does it have a *standard* (标准的) development platform and interface, it can also achieve *free and flexible reorganization* (自由、灵活地重组). The core technology of component GIS development tools is Microsoft's Component Object Model (COM).

GIS developers can achieve GIS system development by using the type of development tools if they are familiar with Windows-based integrated development environment and understand the properties, methods, and events of *controls* (控件). They do not have to master the specialized development language. Therefore, the GIS software development technology is indeed a very *efficacious* (有效的) approach in *seamless integration* (无缝集成), transparency and flexibility.

2.3.4 WebGIS Development Tools

WebGIS is an internet platform-based GIS developed to achieve the sharing and interoperability of geographic information. Generally, it is a kind of socialized GIS with good *expandability* (可扩展性) and cross-platform features. Although WebGIS is still in the early development stage, a lot of companies have launched various WebGIS development tools.

2.4 GIS Software Functions

In practical GIS applications, many GIS software are expected to achieve the functions such as collecting, editing, storing, managing, querying, retrieving, processing, analyzing, sharing and, interchanging geographic data (Table 2-1). They are also expected to be updated. Amongst these different functions, spatial processing and analysis is the kernel of GIS.

Table 2-1 Main Elements of GIS Data Processing and Analysis

GIS Data Processing
Editing processing (编辑处理): editing of graphic data and attribute data
Graphic format processing (图形幅面处理): *splicing and segmentation of graphic data*; *screen cut* (图形的拼接与分割、窗口裁剪)
Transform processing (变换处理): transformation of projection, coordinate and scale; geometric correction (投影变换、坐标变换、比例尺变换、几何校正)
Encoding and compression processing (编码和压缩处理): data encoding; *removal of redundant node* (冗余节点的剔除); data compression
Data interpolation(数据的插值): *interpolation/extrapolation* (内插与外推) of point, region, etc.
Data type conversion (数据类型的转换): conversion of vector data and raster data; data format conversion between systems
GIS Spatial Analysis
Query analysis: topological query; conditions query
Geometric analysis (几何分析): calculation of distances, perimeter and area; *windowing analysis* (开窗分析); *merge of polygon* (多边形合并)
Terrain analysis (地形分析): analysis of spatial interpolation, *contour* (等值线) analysis, slope and aspect analysis, *back bone* (分水岭) analysis, *submerging* (淹没) analysis
Watershed analysis (流域分析): *profile* (剖面) analysis, display and analysis of three-dimensional terrain
Overlay analysis (叠置分析): overlay analysis of polygons, complex analysis of *visual* (视觉) information, conditional or non-conditional overlay analysis
Neighborhood analysis (邻域分析): *buffering* (缓冲) analysis, *corridor* (走廊) analysis, *Thiessen polygon* (泰森多边形) analysis, *fitting* (拟合) analysis
Network analysis: *optimal path* (最佳路径) analysis, *spatio-temporal planning* (时空规划) analysis, network traffic simulation analysis
Image analysis: image *enhancement* (增强), image *segmentation* (分割), image *thinning* (细化), spatial *filtering* (滤波)
Multivariate analysis (多元分析): *cluster* (聚类) analysis, principal component analysis, *discriminant factor* (判别因子) analysis, trend surface analysis, *regression* (回归) analysis
Application model analysis: analysis using various applications models which are closely related GIS

2. 5 Types of GIS Software

GIS has grown from its initial commercial beginnings as a simple *off-the-shelf*（现成的）package to a complex of software, hardware, personnel, institutions, networks, and activities. Currently, only one GIS vendor can offer many different products for distinct applications.

2. 5. 1 Desktop GIS

Since the mid-1990s, desktop GIS has gradually been the major GIS implementation and the most widely used category of GIS software.

Compared to general-purpose GIS, desktop GIS contains programs with low GIS functionality. Running on any personal computer, it does not require a special database server. It is generally used for visualization and easy evaluation of GIS data. Because of its simpler interface, it has been found in widespread use in small companies.

For instance, programs in desktop GIS include ESRI ArcView, MapInfo Professional, SICAD Spatial Desktop, Autodesk Map, Intergraph GeoMedia and others.

2. 5. 2 Server/Internet GIS

Currently, GIS has gradually been dominated by Server/Internet GIS products, which enable us to access, share and distribute geodata over the Internet. The GIS application usually runs on a web server, where it can be queried on PC or cell phone equipped with no more than a web browser and sometimes a plugin. A Server/Internet GIS is a GIS that runs on a computer server that can handle concurrent processing requests from a range of networked clients. Its products have the potential for the largest user base and the lowest cost.

Server/Internet GIS has the largest number of users, although typical Internet users focus on simple display and query tasks[尽管互联网用户仅聚焦在(地理数据的)简单显示和查询任务上，但网络 GIS 还是拥有最大数量的用户].

Examples of software would be Intergraph GeoMedia WebMap, Autodesk MapGuide, ESRI ArcGIS Server, GE Spatial Application Server, Mapinfo MapXtreme and others.

2. 5. 3 Mobile GIS

For data acquisition on a handheld mobile device, GIS software assisted by a GPS was developed to support data capture directly from the field. However, in recent years the demand for mapping services and spatial data analysis on mobile devices has been constantly increasing, driving companies to add more functionality to their products. Nonetheless, the functionality of data management is still not very common due to its demand on powerful hardware.

Services could be navigation and route optimization, showing points of interest, or on-site decision support for firefighters or emergency workers.

A very recent trend is the availability of handheld software on high-end or *smart phones*(智能手机). Thus, many systems are designed to work with server GIS products.

Examples of applications would be ESRI ArcPad, Autodesk OnSite, StarPal HGIS, MapInfo MapXtend, Intergraph Intelliwhere.

2.5.4 Developer/Component GIS

With the advent of component-based software development, many GIS applications can be extended due to their modular structure. There are also many *standalone* (独立的) programs to solve specific tasks. A reasonably knowledgeable programmer can create a purpose-specific GIS application with the GIS feature (component) toolkit. (一个普通的程序员能够通过 GIS 功能/组件工具包创建一个特定目的的 GIS 应用), which interests him because such components can be used to create highly customized and optimized applications. These application programs can either stand alone or be embedded in other software systems.

Examples ofcomponent GIS products include ESRI MapObjects and ArcGIS Engine, ENVI Plugin IDL, MapInfo MapX, and others.

2.5.5 Open Source GIS

Open source applications are programs for free use and on open standards. Often, they are developed by voluntary programmers or academic institutions. They are published under the GNU General Public License (GPL, 通用公共许可证) and permit users to copy, change and redistribute programs together with their source code. Users are authorized to adapt the application to their own specific needs.

There are many different applications, such as GRASS, MapServer, GeoServer.

2.5.6 Others

Based on CAD programs, CAD-GIS applications are ideal for drawing and constructing in engineering. However, they are weak in data management, spatial analysis, cartography, data integration and sharing. Examples of CAD-GIS products include Autodesk Map 3D, CADdy Vermessung and MicroStation GeoGraphics.

DB Server-GIS: Many enterprise-wide GIS incorporate middleware (middle tier) GIS data and application servers to manage multiple users accessing continuous geographic databases, which are stored and managed in *commercial-off-the-shelf* (COTS, 商业现成的) database management systems (DBMS). GIS middleware products offer centralized management of data, the ability to process data on a server, and control over database editing and update. A number of GIS vendors have developed technology services that fulfill this function. Examples of GIS application servers include Autodesk GIS Design Server, ESRI ArcSDE, and MapInfo SpatialWare.

Raster-based GIS, as the name suggests, focuses primarily on raster (image) data and raster analysis. Just as many vector-based systems have raster analysis extensions (for example,

ESRI ArcGIS has Spatial Analyst. and GeoMedia has Image and Grid). In recent years, raster systems have added vector capabilities (Leica Geosystems ERDAS IMAGINE and Clark Labs' Idrisi). In recent years, raster-based systems have added vector capabilities, and now have vector capabilities built in). The distinction between raster-based and other software system categories is becoming increasingly *blurred* (难辨的) as a consequence. The users of raster-based GIS are primarily interested in working with imagery and undertaking spatial analysis and modeling activities.

2.6 Main Vendors of GIS Software

(1) ESRI Inc.

Environmental Systems Research Institute — ESRI Inc. in Redlands, California, a major software developer and vendor today, was founded in 1969 by *Jack Dangermond*. As a typical high-end GIS vendor, it has a wide range of mainstream products which grab the main technical and industry markets. It is a technically-led geographic company focused squarely on the needs of hard-core GIS users.

ArcGIS is the overall name of the ESRI GIS software products. ArcGIS provides a standard-based platform for spatial analysis, data management, and mapping. It can beapplied via the Web, desktop applications, and mobile devices. ArcGIS can be integrated with work order management, business intelligence and other enterprise systems.

(2) Intergraph Inc.

Like ESRI, Intergraph Inc. is also the leading global provider of engineering and geospatial software that enables customers to visualize complex data. Intergraph Inc. was also founded in 1969 as a private company. The initial focus from their Huntsville, Alabama offices was the development of computer graphics systems. After going public in 1981, Intergraph grew rapidly and diversified into a range of graphics areas including CAD and mapping software, consulting services and hardware. After a series of reorganizations in the late 1990s and early 2000s, Intergraph is today sectioned into four main operating units: process, power and marine; public safety; solutions; and mapping and geospatial solutions. The last is the main GIS focus of the company.

Intergraph has a large and diverse product line. From a GIS perspective the principal product family is GeoMedia which spans the desktop and network (Internet) server markets.

(3) Autodesk Inc.

Autodesk, founded in 1982, is now a large and well-known publicly traded company with the headquarter in San Rafael, California. It is one of the world's leading companies serving digital design and content. In detail, services include building, manufacturing, infrastructure, digital media, and location services. Autodesk is best known for its AutoCAD product family which is used worldwide by more than 4 million customers.

Autodesk's main product areas: desktop, where Autodesk Map 3D (based on AutoCAD) is the flagship; an Internet server called MapGuide; and hand-held GIS.

(4) SuperMap Co., Ltd.

SuperMap Software Co., Ltd. (超图) was founded in 1997 by a group of GIS experts and researchers dedicated to developing and providing the most innovative GIS platform software for the worldwide customers. SuperMap is headquartered in Beijing, China. Through years of efforts, SuperMap has built an international distributor network in dozens of countries and developed a series of innovative GIS products, including Desktop GIS, Service GIS, Component GIS and Mobile GIS, which makes SuperMap GIS known as one of the most complete GIS software platforms.

(5) Zondy Cyber Group

Wuhan Zondy Cyber Science and Technology Co., Ltd. (中地数码) tightly associated with China University of Geosciences in Wuhan and GIS Software and Application Engineering Research Center of Ministry of Education is dedicated to creating a new digitalized living environment for all human beings. From MapCAD launched in 1991 to MapGIS in 1995, Zondy Cyber's innovation in the past two decades has been a solid foundation for the information technology development. Nowadays, MapGIS is providing supports to government departments, disaster response, transportation, facility management, environmental management and protection, electronic navigation, business applications, public services and so on.

(6) GeoStar Co., Ltd.

Founded in 1993 by an elite team of developers, Wuda Geoinformatics Co., Ltd. (GeoStar, 吉奥之星) has developed the first GIS platform in China with full intellectual property rights — GeoStar©, which has been widely adopted in various industries related to national economy and social construction projects.

(7) Lingtu Technology Co., Ltd.

Founded in 1999, Beijing Lingtu Technology Co., Ltd. is headquartered in Beijing, China. It is a solution provider with spatial information data products and services as its base. Backed bystate-of-the-art geographical information, communications and GPS technology, it focuses on Location-Based Service (LBS, 位置信息服务), Digital City development, smart traffic and social emergency response system development and convenient map service for the public.

(8) KQ GEO Technologies Co., Ltd

KQ GEO Technologies Co., Ltd. (苍穹数码) was incorporated in Beijing Economic and Technological Development Zone in May 2001. With over ten years' development, the company has become a high-tech private enterprise that integrates research and development of software and hardware as well as surveying and mapping in 3S field. KQ GEO focuses on the development of core technologies and products such as remote-sensing, geoinformation, satellite navigation etc., and its businesses cover data collection and processing, research and development of GIS platform and remote-sensing platform, development and production of satellite navigation software and hardware, informatization solutions, public applications and services etc. With years' growth in

3D field, the company has developed a lot of core technologies, and self-developed software and hardware platforms such as KQGIS Desktop, KQGIS Server, KQGIS Mobile, Multisource RS Data Integrated Processing System, High-accuracy RTK Field Mapping System etc.

Vocabulary

advent[ˈædvent]　*n.* 出现；到来
available[əˈveɪləbl]　*adj.* 可利用的；可得到的；有空的；有效的
architecture[ˈɑːrkɪtektʃər]　*n.* 建筑学；结构；一座建筑物；总称建筑物；建筑风格
component[kəmˈpoʊnənt]　*n.* 组成部分；成分　*adj.* 组成的；构成的
configuration[kənˌfɪɡjəˈreɪʃn]　*n.* 结构；布局；形态
conversion[kənˈvɜːrʒn]　*n.* 转变；换算
client[ˈklaɪənt]　*n.* 委托人；客户
concurrent[kənˈkərənt]　*adj.* 同时发生的
construction[kənˈstrʌkʃn]　*n.* 建设；结构；建筑物；建造；构造
evolution[ˌiːvəˈluːʃn]　*n.* 发展；演变
enterprise[ˈentərpraɪz]　*n.* 企业；事业；谋划
geospatial[dʒiːəsˈpeɪʃl]　*n.* 地球空间信息
geocode[dʒiːəʊˈkəʊd]　*n.* 地理编码
integrate[ˈɪntɪɡreɪt]　*v.* 整合；结合；取消隔离
interface[ˈɪntərfeɪs]　*n.* 界面；接口
infrastructure[ˈɪnfrəstrʌktʃər]　*n.* 基础；基础设施
integration[ˌɪntɪˈɡreɪʃn]　*n.* 集成；综合；同化
incorporate (Inc.)[ɪnˈkɔːrpəreɪt]　*v.* 合并　*adj.* 合并的
maintain[meɪnˈteɪn]　*vt.* 维持；坚持；断言
multiple[ˈmʌltɪpl]　*adj.* 多种多样的；许多的
metadata[ˈmetədeɪtə]　*n.* [计]元数据
mainstream[ˈmeɪnstriːm]　*n.* 主流　*adj.* 主流的
military[ˈmɪləteri]　*adj.* 军事的
overlay[oʊvərˈleɪ]　*v.* 覆盖；铺…上面　*n.* 覆盖物；重叠
platform [ˈplætfɔːrm]　*n.* 平台；站台
private[ˈpraɪvət]　*adj.* 私人的；个人的；私下的；私有的；缄默的
potential[pəˈtenʃl]　*adj.* 潜在的；可能的
procedure[prəˈsiːdʒər]　*n.* 程序；手续；步骤
perspective [pərˈspektɪv]　*n.* 远景；看法；透视　*adj.* 透视的
staff[stæf]　*n.* 全体人员；拐杖；杆　*vt.* 配备员工
significant[sɪɡˈnɪfɪkənt]　*adj.* 重要的；有意义的；意味深长的；显著的
simulation[ˌsɪmjuˈleɪʃn]　*n.* 模拟；仿真；赝品
vendor[ˈvendər]　*n.* 自动售货机；小贩；卖方；供货商

variety [vəˈraɪəti] *n.* 多样；种类；多样化
visualization [ˌvɪʒuəlaɪˈzeɪʃn] *n.* 可视化；形象化

Questions for Further Study

1. Explore the evolution and development of GIS software with the characteristics in different period.

2. Discuss the role of each of the three tiers of software architecture in an enterprise GIS implementation.

3. Evaluate the different types of GIS software systems that might be implemented to fulfill your needs.

4. Have a general overview of GIS functions; try to think about how to allocate these functions in a distributed system when we go through all the functions.

5. Check the websites of the main GIS software vendorsand have an introduction to their commercial products.

Chapter 3

Representing the Real World

The geographic world is extremely complex. The closer we look, the more detail we'll be able to describe. To represent any part of it, it is necessary to choose what in details to be represented over what period of time(地理世界是极其复杂的，我们看得越近，就会对它描述得越详细。为了对地理世界的任一部分进行表达，就必需对在什么样的细节度上及在什么时间段内表达什么内容等做出选择).

Representations have many purposes because they allow us to learn, think, and reason about spatial and temporal factors beyond our first-hand experience, which is the basis of scientific research and planning, and multiple forms of daily problem solving(表达有许多用途，因为表达能够让我们学习、思考，对我们直接经验之外的位置和时间进行推理。这是科学研究、规划的基础，也是解决日常问题的多种形式).

3.1 Geographical Representation

3.1.1 Necessity of Representation

Of the approximately 500 million square kilometers of the earth's surface, only one fifth is land. A fraction of the land is covered by cities and towns where most of us live. The rest of the earth including the parts we have never visited, the atmosphere, and solid ground under our feet, remains unknown to us although these information has been communicated to us through books, newspapers, TV, the webs, or the spoken word. The world is infinitely complex and we only know a few things about the earth because we have observed so little about our Earth directly. We rely on a host of methods for learning about its other parts, for deciding where to go as tourists or shoppers, choosing where to live, running the operations of corporations, agencies, and governments, and many other activities.

Almost all human activities at any time require knowledge about the earth because they occur spatially and temporally. It has been so in the past, and so it will be in the future(目前，在任何时间内，几乎所有的人类活动都需要了解与地球相关的知识，过去如此，将来也如是，因为人类活动既发生在空间上，也发生在时间上).

In order to serve a variety of purposes such as planning, resource management and conservation, travel, the day-to-day operations of parcel delivery service, etc. Humans need to assemble far more knowledge about the earth than we ever did as individuals. And it is necessary to build representations, which are *reinforced* (增强) by the rules and laws that humans have learned to apply to the unobserved world around us.

3.1.2 Geographical Representation

Geographic representation is concerned with the earth's surface or near-surface(地理表达只关乎地球表面或近地表面). For example, in *the Age of Discovery* (地理大发现时代，又称探索时代或大航海时代，from the 15th century to the 18th century), maps became extremely valuable digital representations of geographic knowledge. Representation is the core of problem solving with digital tools (表达是我们用数字工具解决问题的核心所在). There is *a multitude of* (许多) possible ways of representing the geographic world in digital form, however, none of them is perfect for all applications. The key GIS representation issues are what to represent and how to represent it. One of the most important criteria for the usefulness of a representation is its accuracy. Because of infinite complexity of the geographic world, choices always need to be made on building any representation—what to include, and what to leave out.

Geographic data is associated with location, time, and attributes (地理数据与位置、时间及属性有关). A place, a time, or some *descriptive* (描述性) property is a very important element for a geographic datum. In general, a geographic representation of some part of the real world will allow us to enter and store data, analyze the data and then transfer it to humans or other systems. Common geographical representations are as follows.

(1) Maps

A map is a typical geographical representation (地图是一种典型的地理表达方式), which has been used for thousands of years to represent information about the real world and has proven to be extremely useful for many applications in various *domains* (领域). The conception and design of map have developed into a science with a high degree of *sophistication* (精密，复杂).

Paper maps are *best-known* (流传久远的) and used in our lives. The disadvantage of maps is that *are restricted to two-dimensional static representations* (局限于二维静态表达), and that that are always displayed *in a fixed scale* (固定的比例). The map scale determines the spatial resolution of the graphic feature representation at a certain *level of detail* (细节度). The selection of a proper map scale is the first and most important step in map design (合适的地图比例选择是地图设计的第一步，也是非常重要的一步).

With the development of computer networking technology, digital maps—a kind of *analogue cartography* (digital cartography，模拟地图或数字地图)—has enter our daily lives. The role of the traditional maps has changed accordingly. Increasingly, data storage by maps is being taken over by (spatial) databases.

(2) Modelling

Modelling is used in many different ways and has different meanings (建模被用于不同的方面且具有多种不同的含义). A geographical representation of the real world can be considered as a model of the world because the representation has certain characteristics *in common with* (与…有相同之处) the real world. This allows us to build and study the model instead of the real world. The advantage of this is that we can deal with the model and look at different scenarios easily (这样处理的好处是我们可以比较容易地研究模型并可以不同的情景去观察). Therefore, we often change the data in the model to see how the results of the model change. In the GIS environment, the most familiar model is a map (在 GIS 环境中，最熟悉的模型就是地图).

Representation is, therefore, about the choices made in capturing knowledge about the world. The real world can only be described in terms of conceptual models, then real world observations are programmatically translated into meaningful data in GIS. The process of interpreting reality by referring to the real world and a data model is called mathematical modeling (将现实世界和数据模型结合起来解释现实事物的方法叫作数学建模).

Representing the "real world" in a data model has been a challenge for GIS since their inception in the 1960s (用数据模型来表达现实世界是 20 世纪 60 年代 GIS 概念提出以来最大的挑战). A GIS data model enables a computer to represent real geographical elements as graphical elements (GIS 数据模型可以使计算机用图形要素来表达现实的地理要素). We use these GIS data models to simplify the diverse world as well as understand it. However, since we perceive the world differently and have different purposes of using a GIS, it is a practical choice to adopt existing data models and implement a standardized approach.

It is important to adopt a particular model, whether the user realizes it consciously or not. This will influence the data type that may be used to describe the phenomena and the spatial analysis that may be undertaken.

(3) (Spatial) Databases

A database is a *repository* (仓库) capable of storing large amounts of data. Databases can store almost any *sort* (种类) of data. Modern database systems organize the stored data in *tabular format* (列表方式). A database may have many tables, each of which stores data of a certain kind. It is not uncommon that a table has thousands of data rows, sometimes even hundreds of thousands.

The essential function of the spatial data stored and manipulated is to subdivide the earth's surface into entities or objects that can be described (我们存储与使用的空间数据的最基本的功能就是把地球表面划分成我们可以描述的对象或实体). In this way, the content of a spatial database is a model of the earth, and it is a specific type of database. The database stores representations of geographic phenomena in the real world . It is special in the sense that it uses *other techniques than tables* (表格之外的其他技术) to store these representations. This is because it is not easy to represent geographic phenomena using tables.

The spatial database is not the same thing as a GIS, though they have a number of common-characteristics. A spatial database has the functions of databases in general: *concurrency* (并发性), storage, integrity, querying, and in particular spatial characteristics. A GIS, on the other hand, focuses on operating on spatial data. It is known for spatial reference systems and functionalities like distance and area computation, spatial interpolations, digital elevation models et al in geographic space (在地理空间里有空间参考系统以及诸如距离、面积计算、空间关系、数字高程模型等功能). Obviously, a GIS must also store its data, and for this it provids relatively rudimentary facilities(显然，一个 GIS 系统必须存储数据，为此 GIS 提供了一个相对基础的工具). The GIS is increasingly seen to make patial analysis in GIS applications and have a separate spatial database for the data storage (我们越来越多的看到，很多 GIS 应用中都利用 GIS 系统来进行空间分析，并且，数据的存储是在一个独立的空间数据库中完成的).

3.2 Perception of the Real World

In many ways, GIS presents a simplified view of the real world. Irregular and constantly changing, the real world can be described as countless phenomena, from basic *particles* (微粒), the dimensions of oceans and continents, even to the whole earth, the human *perceptions* (感知) of the real world are seldom *identical* (同一的) due to different observers. For example, a surveyor might see a road as two edges to be surveyed, the roadwork authority might regard it as an *concrete surface* (混凝土路面) to be maintained, and the driver sees it as a highway. Our perception of the real world would be quite different if we were viewing the scene from the top of a hill or building, or if we had walked, driven, or flown over the environment. It would also be different if we were describing the scene captured by photographs or images. Our living environment, cultural background and acknowledge on the real world influence our interpretation of the the features observed and of what you decide to ignore. The complexity of the real world and the broad spectrum of its interpretation suggest that GIS system designs will vary according to the capabilities and preferences of their creators (与表现出解释的广谱性一样，自然界的复杂性预示着 GIS 的系统设计会根据开发者的认知能力、喜好而改变). Meanwhile, the world is infinitely complex, but the computer system is finite. Representations in GIS must somehow limit the amount of details captured (同时，自然界是非常复杂的，但计算机系统却是有限的，因此 GIS 中的表达必须适当限制所收集的细节量). The methods of generalization are used to remove details unnecessary for an application in order to reduce data volume and speed up operations (很多概括的方法用来剔除那些不必要的细节以减少数据量并加快运算速度).

One common way of limiting details is to throw away information that applies only to small areas. In other words, it is observed far away from earth surface. Sometimes we humans want the information of real spatial objects (such as roads, land use, elevation, trees, waterways, etc.), without any other irrelevant information. In fact, real spatial objects can be divided into two abstractions: the one is *discrete objects* and the other is *continuous fields*(现实世界的空间实体可以抽象两类：一类是离散的对象，另一类是连续的场). For example, a house in the countryside

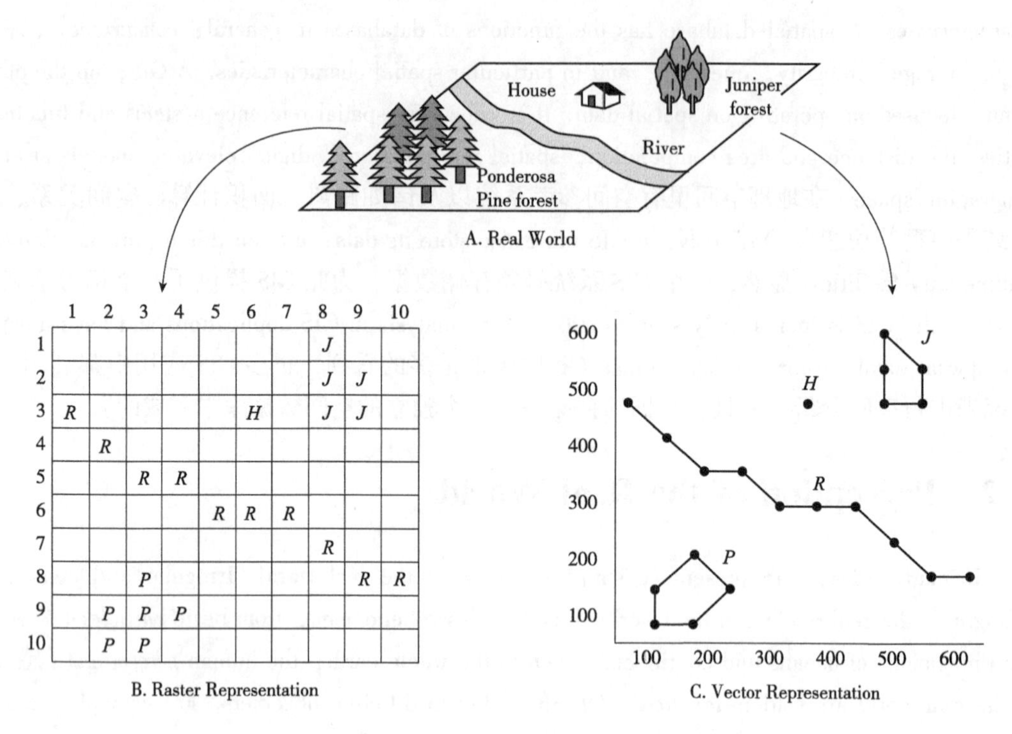

Figure 3-1 Two Methods of Representing the Real World

is conceived as a discrete object with its clear boundaries and empty space surrounded (Figure 3-1).

3.2.1 Discrete Object

The discrete objectis known as one way of representing the geographic world (离散对象是表达地理世界的方法之一). In this view, the world is empty, and it is occupied by objects with *well-defined boundaries* (明确的界限) that are instances of *generally recognized categories* (是通常具有公认类别的实例). Geographic objects are identified by their *dimensionality* (维度). Objects that occupy an area are termed two-dimensional (2D), and generally referred to as region (占据一定面积的称为二维, 通常称为区域). The term polygon is also common for technical reasons. Other objects are often represented as one-dimensional objects such as roads, or rivers, and generally referred to as lines. Other objects are more like zero-dimensional points, such as individual buildings or trees, and are referred to as points. All of this information can be expressed in a table with each row corresponding to a different discrete object, and each column to an attribute of the object. To reinforce a point made earlier, this is a very efficient way of capturing raw geographic information on discrete objects.

The discrete object view leads to a powerful way of representing geographic information about discrete objects. But it is not perfect as a representative for all geographic phenomena(离散对象的观点是一个对物体非常好的地理信息表达方法, 但是这种方法不是对所有地理现象表达都很完美). Imagining when we observe the earth from another planet, what can be used to represent the complexity of the real world and environment? It is hard for a table to impress us with natural phenomena such as rivers, landscapes, or oceans(想象一下，我们从另一个星球来观察地球，选择一个什么样的方法来表达人类周围复杂的环境。用一个表格的形式很难表达出给人以突出印象的自然现象，如河流、陆地景观或海洋等). The properties of a river or an ocean is hardly represented by a table in a clear manner. So, the discrete object view works well for some kinds of phenomena, but it fails in other aspects(地球上没有什么东西看来像一个表格。用一个表格来表达河流或者海洋的性质一点也不清楚。虽然离散对象的方法对一些现象表达很好，但对另外一些要素来说也丢失了一些信息).

3.2.2 Continuous Field

The discrete object view leads to a powerful way of representing geographic information about objects (离散对象的方法是一种非常有效的地理世界的表示方法). While we might think terrain as discrete mountain peaks, valleys, ridges, slopes, etc. and list them in tables and record in numbers, unresolvable problems of representing these objects will occur(当我们用离散的山峰、山谷、山脊、坡度等用列表并计数描述地形时，对所有这些离散的对象就出现了许多不可解决的问题). Instead, it is much more useful to think terrain as a continuous surface, in which *elevation* can be defined *rigorously* (严格的) at every point. Such a continuous surface has formed the basis of representing other geographic phenomena, known as the continuous field view (这样连续的表面形成了其他地理现象表达的基础，如连续场观). In this view the geographic world can be described by a number of *variables* (变量), each measurable at any point on the earth's surface, and changing in value across the surface.

The continuous field view represents the real world as a finite number of variables, each one defined at every possible position(连续场观点表示现实世界为一个有限的变量，每个变量在每个可能的位置上都很明确).

Objects are distinguished by their dimensions, and naturally fall into categories of points, lines, or planes(对象是通过其维数来区分的，自然地就分为点、线、面). Continuous fields, on the other hand, can be distinguished by the content and smoothness of the variation. For example, population density is a kind of continuous field. As the number of people per unit area, it can be defined everywhere. Continuous fields can also be classified by land to form land use types or soil types(连续场也可以由土地分类形成土地利用类型或土壤类型). Such fields change suddenly at the boundaries between different classes. Other types of fields can be defined by continuous variation along lines, rather than across space. Continuous fields can be distinguished by what is being measured at each point (连续场可以用测量每个点值来区分). A vector field assigns two variables (magnitude and direction) at every point in space, and is used to represent flow phenomena such as winds or *currents* (电流); fields with only one variable are termed *scalar fields* (标量场).

Continuous fields and discrete objects define two conceptual views of geographic phenomena, but they do not solve the problem of digital representation. Two digital representational ways are possible: *raster* (grid-based) and *vector* (line-based) (Figure 3-1). Raster and vector are two methods of representing geographic data in digital computers(连续场与离散对象定义了两种地理现象的表达方法，但都不能解决数字化的表达的问题。两种数字化的表达方式是可行的：基于格网的栅格和基于线划的矢量。栅格与矢量是计算机表达地理世界的两种方法).

3.3 Vector Representation

The concept assumes that space is discrete rather than continuous and consists of an infinite (in theory) set of coordinates[矢量的表达方式假设地理空间是离散的，而不是连续的，且(在理论上)由无数坐标对组成]. A vector representation is composed of three main elements: *points*, *lines* and *polygons*. That is, the real world can be represented by a series of points (such as towers, telegraph poles), lines (such as roads and highways) and polygons (such as lakes). In other words, a real-world entity could be represented by different types of vector features depending on the map scales used in an application[矢量表达主要由点、线、面三种主要的要素组成。也就是说，现实世界可以用一系列的点(如塔，电线杆)、线(如道路和高速公路)、面(如湖泊)来表达。换句话说，真实世界的实体可以用不同类型的依赖于应用中的地图比例尺的大小矢量要素来表示].

Vector data provides a way to represent the real-world features within the GIS environment. A feature is anything you can see on the landscape. Imagine looking down from the top of a hill, we can see houses, roads, trees, rivers, and so on. Each one of these things would be a feature when we represent them in a GIS application. Vector features have attributes which describe the features with textual or numerical information.

A vector feature is a geometric representation of its shape. The geometry is made up of one or more interconnected vertices (矢量要素是用几何图形来表达其形状的。几何图形是由一个或多个相互连通的顶点组成的). A vertex describes a position in space using an *X*, *Y* and optionally *Z* axis. Geometries which have vertices with a z axis are often referred to as 2.5D since they describe height or depth at each vertex, but not both (具有 *Z* 轴结点的几何图形经常是指2.5维，因为 *Z* 轴结点描述的是高度或深度信息，但不能二者都描述).

When a feature's geometry consists of only a single vertex, it is referred to as point feature. Where the geometry consists of two or more vertices and the first vertex and the last vertex are not equal, a polyline feature is formed. Where four or more vertices are present, and the last vertex is equal to the first, an enclosed polygon feature is formed (Figure 3-2).

Points, lines, and polygons can be used to describe the more complex objects by capturing the internal structure of an entity. An important addition is the inclusion of topology information. Topological relation is independent of the coordinate system and establishes more complex relationships between adjacent entities.

In recent years, object-oriented approach has been applied in database or spatial database, which is a more highly structured way of encapsulating entity data. Object-orientation is described more fully in the context of data structures.

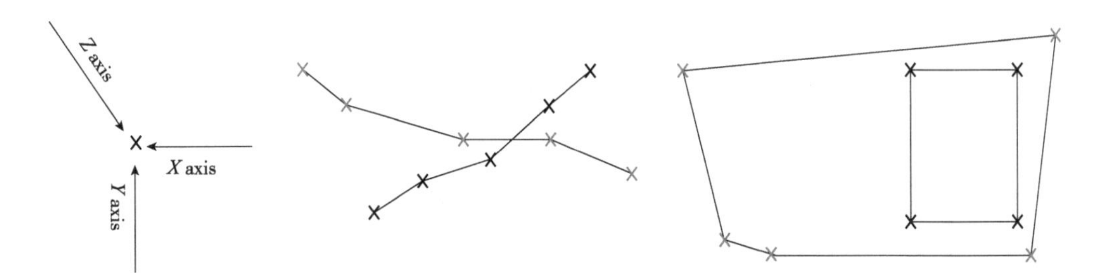

Figure 3-2 Point Feature, Line Feature and Polygon Feature

3.4 Raster Representation

Raster is a *matrix*（矩阵）made up of pixels (also called cells). Each cell contains a value representing the conditions of the area covered by it (Figure 3-3). Based on a cellular organization, raster divides space into a series of units. In a raster representation, space is divided into an array of rectangular (usually square) cells that we call grid cell（栅格表达是将空间分成一系列我们称之为格网单元的矩形或正方形单元）, and all geographic variables are then expressed by assigning properties or attributes to these cells（通过给这些单元指定属性来表达地理变量）. Grid cells are the most common raster representation. Spatial features are divided into grid cells, and a coordinate (X, Y) is assigned to each cell, as well as a value. In general, raster representations divide the world into cellular arrays and assigned attributes（通常，栅格表达将自然界分成系列单元以及其指定的属性）.

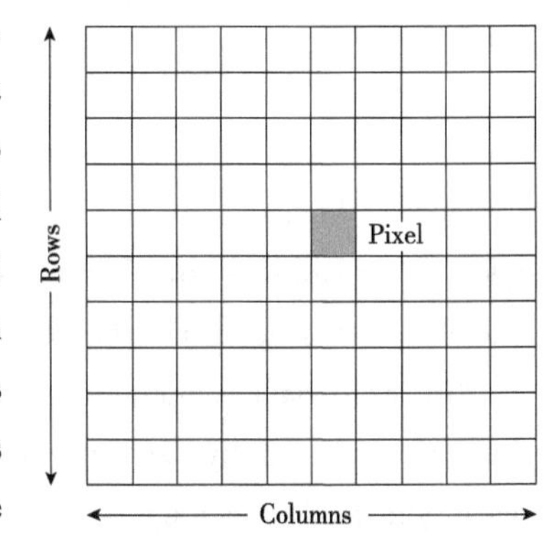

Figure 3-3 Raster

Raster data is used in a GIS application when we want to display information that is continuous within an area and cannot easily be divided into vector features. Point, polyline and polygon features work well for representing some features on this landscape, such as trees, roads and building footprints. But some other features on a landscape can be more difficult to be represented by using vector features. For example, the grasslands have many variations in color and *density*（密度）of cover. It would be easy enough to make a single polygon around each grassland area, but a lot of the information about the grassland would be lost in the process of simplifying the features to a single polygon. When attribute values are assigned to a vector feature, they apply to the whole feature, so vectors aren't very good at representing features that are not *homogeneous*（均匀的，同种的）all over. Another approach you could take is to digitize every small variation of grass color and cover as a separate polygon. The problem with that approach is that it will take a huge amount of work in order to create a good vector dataset. Using raster data is a solution to these problems. Raster data is not only good for images that *depict*（描述）the real world surface,

e. g. satellite images and *aerial photographs* (航空相片), but also for representing more abstract ideas. For example, raster can be used to show rainfall trends over an area, or to depict the fire risk on a landscape. In these kinds of applications, each cell in the raster represents a different value.

A raster representation also relies on tessellation: geometric shapes that can completely cover an area. Although many shapes are possible[e. g. *triangles* (三角形) and *hexagons* (六边形)], squares are the most commonly used. Every raster layer in a GIS has pixels (cells) of a fixed size that determine its spatial resolution. This becomes apparent when you look at an image at a small scale and then zoom in to a large scale. Resolution is an important concern in raster representations.

Several factors determine the spatial resolution of an image. For remote sensing data, spatial resolution is usually determined by the capabilities of the *sensor* (传感器) used to take an image. For example, SPOT 5 satellites can take images at 10 m×10 m per pixel. Other satellites, such as MODIS (中分辨率光谱成像仪) take images only at 500 m×500 m per pixel. In aerial photography, 50cm×50cm pixels are not uncommon. Images with a pixel size covering a small area are called "high resolution" images because they display a high degree of detail and clarity. Images with a pixel size covering a large area are called "low resolution" images because they display a low degree of detail and clarity.

In raster data computed by spatial analysis, the spatial density of information used to create the raster will usually determine the spatial resolution. For example, if you want to create a high-resolution map on average rainfall, you will ideally need many *weather stations* (气象站) in close proximity to each other.

To capture raster at a high spatial resolution, one of the main things is storage requirement. Think of a raster with 3×3 pixels, each of which contains a number representing average rainfall. To store all the information contained in the raster, you will need to store 9 numbers in the computer's memory. Now imagine you want to have a raster layer for the whole of China with pixels of 1 km×1 km. It means your computer would need to store over a million numbers on its hard disk in order to hold all of the information. Making the pixel size smaller would greatly increase the amount of storage needed. Sometimes using a low spatial resolution is useful when you want to work on a large area and are not interested in looking at any one area in a lot of details.

The cloud map you see on the weather report, is an example of this—it's useful to see the clouds across the whole country. Zooming in to one particular cloud in high resolution will not tell you very much about the *upcoming weather* (未来天气). On the other hand, using low resolution raster data can be *problematic* (有疑问的) if you are interested in a small region because you probably won't be able to *make out* (理解, 辩认出) any individual features from the image.

Just as there are numerous varieties of organizing *geometric* (几何的) and *topological* (拓扑的) properties in data tables, so there are different ways of *structuring regular tessellation data* (构建规则格网数据). The common form for structuring regular *grid cells* (格网单元) consists of a row coordinate value, a column coordinate value, and possibly a cell *identifier* (标识符). The origin of rows and columns is typically at the upper-left corner of the raster. *Multi-valued* (多值的) raster can be constructed by the combination of single valued raster or by the combination of multi-valued raster.

The records of per-grid cells can be listed sequentially in row order or column order. They can also be arrayed in the computer storage format as the entire matrix of cells(每个格网单元的记录可以在行或列按顺序地列出，或者是利用计算机存储矩阵数据的格式列出). The matrix format may also be used if there are several data items for each grid cell(如果格网单元中有几个数据项，则也用矩阵). Although often associated with a map layer concept, the overlay arrangement is not a requirement for regular tessellationsas the grid-cell or pixel form suggests. However, the matrix and clustered forms are designed to work on thematic layers, meaning that the attribute data are not recorded in sequence for each and every cell.

For entity-oriented representations, attributes are separated from the spatial information in most cases, whereas for regular tessellations the positional and attribute data are associated (对于有方向的实体的表达，大多数情况下空间信息与属性是分开的，而一般规则格网数据位置与属性是关联的). Regular cells are often used to aggregate or disaggregate things across scales (规则单元常用于跨尺度的聚合或分解的事物). The resultant form, known as the pyramid model, provides a multiple scale representation, with spatial units constant for a given scale (金字塔模型的结果形式，提供了一种多尺度表达，对于给定的尺度，空间单元大小不变). When information is represented in raster form, all details about variation within cells are lost, and instead the cell is given a single value.

3.5 Attribute Data

3.5.1 Concept

GIS data sets include not only spatial data (such as vector data and raster data), but also attribute data that can be linked to the spatial data. Attribute data are used to represent the presence or lack of a certain characteristic about the spatial entity. Attribute data in GIS are stored in tables. And attribute tables are organized by rows and columns. Each row represents a spatial feature, and each column represents a property or characteristic of the spatial entity. According to the human perception of the real world, attribute data can also be continuous or discrete.

Theoretically, continuous data have an infinite number of measurements depending on the resolution of the measurement system. There are no limits to the gaps between the measurements. It is data that can be expressed on an infinitely divisible scale. Even if the measurements range from 0 to 1 there may be an infinite number of measurements within (0.000000000000... to 0.999999999999...). The continuous random variables can be any of the infinite number of values over a given interval (间隔). These variables generally represent things measured, not counted. For example: air temperature, surface height, air pressure, coverage, etc. Continuous data must *be converted to*（被转换）a form of *variable data*（变量数据）called discrete data in order to be counted or useful.

Discrete data is another type of data with a finite number of measurements and based on counts. Data can be sorted into distinct, countable, and in completely separate categories. The count cannot be further divided on a meaningful infinite scale. For example: gender, blood type, rating 1-100 (from lowest to highest), *nutrient content*（养分含量）, number of population to a *designated community*（指定的小区）, etc. Discrete data cannot be added to or subtracted from other attribute data.

Continuous data is more precise and informative than discrete data. Continuous data can *remove estimation and rounding of measurements*（去除测量值的估计和四舍五入）, and it often cost much (time, money) to obtain.

3.5.2 Attribute Data Classification

It is necessary to classify attribute data to ensure the correct statistical tools which are used to conduct spatial analysis. Attribute data can be classified into several different types of data, including *nominal*, *ordinal*, *interval* and *ratio*（名义的，顺序的，间隔的和比率的）.

(1) Nominal Data

According to the Dictionary of Statistics and Methodology by *W. Paul Vogt*, a nominal variable is a data type that distinguishes entities by a limited number of classifications including type or category[名义变量是一种将实体通过有限数量的分类(包括类型或种类)来进行区分的数据类型]. Of these different attribute data types, nominal data is of the lowest level and the least numeric categorization. That is, the nominal data can be a numerical label that represents a *qualitative*（定性的）description or numbers that represent a category or classification. Nominal data describe different kinds or different categories of data such as soil type or land use types. For example, the gender can be labeled as 1 = Male or 2 = Female; the geographical orientation can be labeled as 1 = East, 2 = South, 3 = West or 4 = North, and such data types that often result in nominal data as religion, *zip code numbers*（邮政编码）, birth dates, telephone numbers, personal ID number, *ethnicities*（民族）, *educational level*（学位）, *position*... The average or *variance*（方差）of the nominal data is meaningless, and there is also no priority or rank based on these numbers. There are limited statistical techniques to analyze this type of data, but *chi-square statistics*（卡方检验）is the most common.

(2) Ordinal Data

Ordinal data is of the higher level than nominal data. Ordinal data differentiate data by a ranking relationship. Just like the name sounds, ordinal data refers to ordered categorical variable. Items on an ordinal scale are set into some kind of order by their position on the scale. In other words, the order of items is often defined by numbers assigned to them to show their relative position. For example, *the average annual per capital income*（年人均收入）can be grouped into different categories as ¥0-5000, ¥5001-10 000, ¥10 001-100 000,..., which then might be coded as 1, 2, 3,...Similar to nominal data, ordinal data is a form of discrete data and qualitative data. Sometimes, ordinal data is also *round* (不确切的) when ranking geographic phenomena, i. e. *the best livable cities* (最佳宜居城市), *the most beautiful scenic spot* (最美景区). *Nonparametric test* (非参数检验) should be applied to analyze it.

(3) Interval Data

Interval data is classified as the next higher level of data than the above two ones. Interval data have known intervals between values. In spite of the same as ordinal data, interval data is measured along a scale in which the intervals are equally split. Interval data can be either continuous or discrete. It is often used in geographic experiments that measure attributes along an *arbitrary* (任意的) path between two *extremes* (极端, 端点). Interval scales not only tell us about order, but also about the value between each item, such as the relative elevation value per 100 meter on the earth's surface; the level of happiness rated from 1 to 10.

(4) Ratio Data

Ratio data are the same as interval data except that ratio data are based on a meaningful, or absolute, zero value. Population densities are an example of ratio data, because a density of 0 is an absolute zero. It ranks the highest level of data measurement, and can be analyzed in more ways than the other three types of data. Ratio data can also be either continuous or discrete. In a ratio scale, numbers can be compared as multiples of one another, i. e. one person can be twice as tall as another person. Importantly, the number zero has meaning. Examples include time, *volume* (体积, 容量), weight, *voltage* (电压), height, pieces/hour...Interval data and ratio data measure quantities and hence are quantitative.

3.6 Data Output & Visualization

3.6.1 Data Output

Although most GIS professionals are becoming more interested in solving analytical puzzles with GIS tools, most *onlookers* (旁观者) are impressed by the high-quality cartographic output from a GIS. As the most effective vision tool in communicating geospatial data, Maps put emphasis on the location or the distribution pattern of the spatial data. Compiled maps are often shown in two ways: ①*Hardcopy*(硬拷贝) by *printer*(打印机) or *plotter*(绘图仪). The first step for hard copy output is to set printer/plotter and paper size, and then preview. To preview allows us to check if the output is exactly the same as what we set. If the preview is not the same as what we set, we need to print the map separately. ②Raster images of the common format transformed for applications in varied systems. The key to the transformed output is to set the correct resolution of raster images. '.mxd' is the default format of GIS software which cannot be used out of software environment. But GIS software is equipped with the function of transforming '.mxd' format to such formats like EMF, BMP, EPS, PDF, JPG, TIF, etc. The vector or raster data transformed can be used in other environment beyond GIS software.

3.6.2 Visualization

Visualization is known as visualization in scientific computing. It is an *algorithm* (算法) that converts symbols or data to visual graphs, which is easy for users to observe the process of simulation and calculation. Only if geospatial information is transformed into digital information, it can be processed by computers. The transformation is the procedure of visualization. It includes several aspects as follows:

①*Visualization of map data* (地图数据的可视化). It is a screen display of map data. Corresponding variables (i.e. shape, size, color, etc.) are chosen to make a visual map with total or divided elements according to classification features of digital map, such as screen maps, paper maps and *lithographic films* (平版制版胶片).

②Visualization of geographic information (地理信息的可视化). It classifies statistic data, experiment data, observation data and geographic survey data according to various mathematical models. Then it is expressed in the form of thematic map with appropriate visual variables, such as statistic graphs or *histograms* (直方图) with different levels and regions. Visualization of geographic information presents the initial meaning of visualization in scientific computing.

③Visualization of spatial analysis (空间分析的可视化). Spatial analysis is an important part of GIS. Itincludes analysis of network, buffer and overlay. The result of analysis is often expressed by thematic map.

Vocabulary

aggregate[ˈæɡrɪɡət] *vi.* 集合；聚集；合计 *vt.* 集合；聚集；合计 *n.* 合计；集合体；总计 *adj.* 聚合的；集合的；合计的

analogue[ˈænəlɔːg] *n.* 类似物；类似情况；对等的人 *adj.* 类似的；相似物的；模拟计算机的

approximately[əˈprɑksɪmətli] *adv.* 大约，近似地；近于

assemble[əˈsɛmbl] *vt.* 集合，聚集；装配；收集 *vi.* 集合，聚集

cartography[kɑrˈtɑgrəfi] *n.* 地图制作，制图；制图学，绘图法

category[ˈkætəgɔri] *n.* 种类，分类；[数]范畴

characteristic[ˌkærəktəˈrɪstɪk] *adj.* 典型的；特有的；表示特性的 *n.* 特征；特性；特色

concurrency[kənˈkɜrənsi; kənˈkʌrənsi] *n.* [计]并发性；同时发生

constantly[ˈkɑnstəntli] *adv.* 不断地；时常地

continuous[kənˈtɪnjʊəs] *adj.* 连续的，持续的；继续的；连绵不断的

coordinate[koˈɔrdɪnet] *n.* 坐标；同等的人或物 *adj.* 并列的；同等的 *vt.* 调整；整合 *vi.* 协调

criteria[kraɪˈtɪrɪə] *n.* 标准，条件(criterion 的复数)

dimensionality[daɪmɛnʃə'næləti] *n.* 维度；幅员；广延
discrete[dɪ'skrit] *adj.* 离散的，不连续的 *n.* 分立元件；独立部件
distinguish[dɪ'stɪŋgwɪʃ] *vt.* 区分；辨别；使杰出，使表现突出 *vi.* 区别，区分；辨别
elevation[ˌɛlɪ'veʃən] *n.* 高地；海拔；提高；崇高；正面图
essential[ɪ'sɛnʃl] *adj.* 基本的；必要的；本质的；精华的 *n.* 本质；要素；要点；必需品
generalization[ˌdʒɛnrələ'zeʃən] *n.* 概括；普遍化；一般化
geometric[ˌdʒiə'mɛtrɪk] *adj.* 几何学的；[数]几何学图形的
graphic['græfɪk] *adj.* 形象的；图表的；绘画似的
hexagon['hɛksəgɑn] *n.* 六角形，六边形 *adj.* 成六角的；成六边的
homogeneous[ˌhomə'dʒinɪəs] *adj.* 均匀的；[数]齐次的；同种的
infinitely['ɪnfɪnətli] *adv.* 无限地；极其
integrity[ɪn'tɛgrəti] *n.* 完整；正直；诚实；廉正
interpret[ɪn'təprɪt] *vt.* 说明；口译 *vi.* 解释；翻译
irregular[ɪ'rɛgjələ] *n.* 不规则物；不合规格的产品 *adj.* 不规则的；无规律的；非正规的；不合法的
manipulate[mə'nɪpjulet] *vt.* 操纵；操作；巧妙地处理；篡改
optimization[ˌɑptəmɪ'zeʃən] *n.* 最佳化，最优化
perception[pə'sɛpʃən] *n.* 知觉；[生理]感觉；看法；洞察力；获取
reinforce[ˌriɪn'fɔrs] *vt.* 加强，加固；强化；补充 *vi.* 求援；得到增援；给予更多的支持 *n.* 加强；加固物；加固材料
repository[rɪ'pɑzə'tɔri] *n.* 贮藏室，仓库；知识库；智囊团
representation['rɛprɪzɛn'teʃən] *n.* 代表；表现；表示法；陈述
resolution[ˌrɛzə'luʃən] *n.* [物]分辨率；决议；解决；决心
scalar['skeɪlə] *adj.* 标量的；数量的；梯状的，分等级的 *n.* [数]标量；[数]数量
scale[skel] *n.* 规模；比例；鳞；刻度；天平；数值范围 *vi.* 衡量；攀登；剥落；生水垢 *vt.* 测量；攀登；刮鳞；依比例决定
scenarios[sɪ'nɛrɪˌo] *n.* 情节；脚本；情景介绍(scenario 的复数)
sequentially[sɪ'kwɛnʃəli] *adv.* 从而；继续地；循序地
spectrum['spɛktrəm] *n.* 光谱；频谱；范围；余象
square[skwɛr] *adj.* 平方的；正方形的；直角的；正直的 *vt.* 使成方形；与…一致 *vi.* 一致；成方形 *n.* 平方；广场；正方形 *adv.* 成直角地
storage['stɔrɪdʒ] *n.* 存储；仓库；贮藏所
subdivide['sʌbdɪvaɪd] *vi.* 细分，再分 *vt.* 把…再分，把…细分
terrain[tə'ren] *n.* [地]地形，地势；领域；地带
tessellation[ˌtɛsɪ'leʃən] *n.* 棋盘形布置；棋盘花纹镶嵌；镶嵌式铺装；镶嵌细工；棋盘花纹
topological[ˌtɑpə'lɑdʒɪkl] *adj.* 拓扑的；[解]局部解剖学的；[地]地志学的
triangles['traɪˌæŋgl] *n.* [数]三角形，三角形态(triangle 的复数形式)
visualization[ˌvɪʒʊəlɪ'zeʃən] *n.* 形象化；清楚地呈现在心

Questions for Further Study

1. What the world looks like? What is geographic representation?

2. What are continuous fields or discrete objects?

3. How to represent phenomena conceived as fields or discrete objects?

4. What do you understand by the terms raster and vector? How would you decide which to use in any specific given project?

5. "No representation of geographic phenomena can ever be perfect" is this true, are there exceptions, and what implications does this statement have for users of GIS?

Chapter 4

Nature of Geographical Data

In the real world, nearly about 70 to 80 percent information is related to geographical data. The nature of geographic data presents the basic structure of geographic data about scale, accuracy and quality (现实世界中，70%~80%的信息都与地理数据有关。地理数据的自然属性表达了与尺度、精度和质量相关的基本地理数据结构).

4.1 Geographical Phenomena

4.1.1 Basic Concept

A phenomenon is a fact, occurrence, or circumstance that is observed to exist/happen or observable within nature (现象是经观察存在/发生、或自然界可见的一种事实、事件或环境).

> A geographical phenomenon is a phenomenon across space (or physical location), which requires two descriptors: what is present, and where it is. It can be refers to the external form or surface features of things in space, and any phenomenon of the earth's surface.

Frequently, similar to the phenomenon, the geographical phenomenon also tends to be *in close proximity* (接近) to another similar geographical phenomenon. For example, the loess distributed at different places in the world, whether it is in Asia, the America or Africa, usually has similar characteristics as *highly porous and vertical joints* (疏松多孔和垂直节理). Hereby, two basic geographical concepts-*similarity and diversity* (相似性与差异性)—come into being.

Some geographical phenomena change wondrously slowly across space and time, i. e. the air temperature, precipitation, day-to-day growth of height of a kid; while the others vary extremely irregularly (一些地理现象随空间和时间变化非常缓慢，如气温、降水量、小孩每天的身高增长量，而另一些地理现象则极不规则地变化着). Everyone would recognize the extreme difference of landscapes between such regions as the Yangtze River, *the Takla Makan Desert* (塔克拉玛干沙漠), and the Great Wall, and many would recognize the more *subtle* (精细的) differences between the *loess tableland* (黄土塬), *loess ridge* (黄土梁), and *loess hill* (*loess knolls*)（黄土峁）（Figure 4-1）. *Heterogeneity*（异质性，多相性）occurs both in the way the landscape looks, and in the way processes act on the landscape. While the *spatial variation*（空间变异）in some processes simply *oscillates*（振荡，摆动）about an average（controlled variation，可控变异），other processes vary ever more the longer they are observed（uncontrolled variation，不可控变异）. As a general rule, spatial data exhibit an increasing range of values, hence increased heterogeneity, with increased distance.

A. Loess Tableland　　B. Loess Ridge　　C. Loess Knolls

Figure 4-1　Different Types of Loess Landform

First Law of Geography（Tobler's First Law）：everything is related to everything else, but near things are more related than distant things（任何事物都是与其他事物相关的，只不过相近的事物更相关）（Waldo Tobler, 1970）.

This property is known asspatial auto-correlation, which can be assessed by knowledge of the degree and nature of spatial heterogeneity, that is, the trend of differences between geographical locations and regions. It is very likely to take two or more factors into consideration for predicting future change of the real world（地理学第一法则又称 Tobler 第一法则：任何事物都彼此相关，只是较近的事物比较远的事物更加相关。这种属性称为空间自相关性，可由空间异质性的程度和性质即地理位置和地区之间彼此差异的趋势来进行评价。非常可能的是，用以预测现实世界在未来变化时，需要将两个或以上因素的信息一起考虑）.

Second Law of Geography（Law of Spatial Heterogeneity, Goodhild's Second Law）：The isolation of space leads to the difference between objects, that is, heterogeneity（空间的隔离，造成了地物之间的差异，即异质性）.

Third Law of Geography（Law of Geographic Similarity）：The more similar the geographical configuration of two points（regions），the more similar the value（process）of the target variable at these two points（regions）（两个点/区域的地理配置越相似，目标变量在这两个点/区域的值/过程越相似）.

4. 1. 2　Spatial Object

Geographical phenomena existing in space can be described by spatial objects. Spatial objects are classified according to their topological dimension, which provides a measure of the way they fill space（空间对象根据它们的拓扑维度来进行分类，这为空间对象填充空间的方式提供

了一种量度).

(1) Point

A point is a spatial object with no length, breadth or depth, and accordingly with dimension 0. Points may be used to indicate spatial occurrences or events, and their spatial patterning. Point pattern analysis is used to identify whether occurrences or events are interrelated.

A point feature has an *X*, *Y* and optionally, *Z* value. The *X* and *Y* values depend on the Coordinate Reference System (CRS) being used. A CRS is a way to accurately describe where a particular place is on the earth's surface. One of the most common reference systems is longitude and latitude. Since we know the earth is not flat, it is often useful to add a *Z* value to a point feature. This describes how high above sea level you are (objects are).

The point pattern is shown as Table 4-1.

Table 4-1 Point Pattern Analysis(2008 Wenchuan Earthquake)

Magnitude	7.9
Date-time	Monday, May 12, 2008 at 06:28:01 UTC
Location	30.98°N, 103.32°E
Depth	19 km (11.8 miles) set by location program
Region	EASTERN SICHUAN, CHINA
Distances	80 km (50 miles) WNW of Chengdu, Sichuan, China 145 km (90 miles) WSW of Mianyang, Sichuan, China 350 km (215 miles) WNW of Chongqing, Chongqing, China
Parameters	NST=357, Nph=357, Dmin=592.1 km, Rmss=1.38 sec, Gp=14°, M-type=moment magnitude (Mw), Version=R
Source	USGS NEIC (WDCS-D)
Event ID	us2008ryan

Point objects include: entity point used to represent an entity, i.e. elevation point, well, building (实体点: 用以表示一个实体, 如高程点、油井、建筑物), text point used to locate the annotation(注记点: 用以定位注记), label point existing within a polygon for identifying attributes(内点: 存在于多边形内, 用于标识多边形的属性), node, especially end node, used to indicate the start or end of a line or arc(结点: 用以表示线段或弧段的起点和终点) and vertex used to indicate the internal points of a line or arc(拐点、节点、中间点: 用以表示线段或弧段的内部点).

(2) Line/Polyline

A line made up of connected points (nodes) of an object has length, but no breadth, depth. Accordingly, it is one- dimensional. Lines areused to represent linear entities such as roads, rivers, *contours* (轮廓线), footpaths, *pipelines* (流水线), and so on, which frequently build together into networks. Also They are also used to measure distances between spatial objects.

A point feature is a single *vertex* (顶点), when two vertices are joined, a line is created.

When more than two vertices are joined, a polyline is formed. With two or more vertices, a polyline has formed a continuous path drawn through each vertex. Sometimes we have special rules for polylines in addition to their basic geometry. For example contour lines may touch [e. g. at a *cliff face* (悬崖壁)] but should never cross over each other. Similarly, polylines used to store a road network should be connected at intersections. In some GIS applications, you can set these special rules for a feature type (e. g. roads) and the GIS will ensure that these polylines always *comply with* (遵守) these rules.

Sometimes, linear objects are too small to be depicted at the proper scale on a map. *Hierarchy* (层次、等级) of lines are used to show the difference in different objects. For example, a thin road might represent a *dirt road* (土路) whereas a thick road represents a highway. If a curved polyline has very large distances between vertices, it may appear *angular* (生硬的、有角的) or *jagged* (锯齿状的), depending on the scale. Because of this, it is important that polylines be digitized (captured into the computer) with distances between vertices that are small enough for the scale at which you want to use the data.

The attributes of a polyline describe its properties or characteristics, such as length, orientation [compass direction of *mineral veins* (矿脉)]. The attributes of a road polyline may describe whether the road surface is with *gravel* (碎石) or *tar* (柏油), how many *lanes* (车道) it has, whether it is a one-way street, and so on. GIS can use these attributes to symbolize the polyline feature with a suitable color or *line style* (线型).

(3) Area/Polygon

An area/polygon object has dimensions of length and breadth, but not depth. It may be used to represent enclosed areas of natural objects, such as *administrative boundary* (行政界线). Also an area is commonly used to represent *artificially demarcated area* (人为划定的区域), such as *census tracts* (人口普查区).

Polygon features are enclosed areas that can be made up of a circuit of line segments like islands, country boundaries and so on. Like polyline features, polygons are created from a series of vertices that are connected with a continuous line. However, the first and the last vertices should always be at the same place because a polygon always describes an enclosed area. Polygons often have shared geometry-boundaries that are in common with a *neighboring polygon* (相邻多边形). Numerous GIS applications have the capability to ensure that the boundaries of neighboring polygons exactly *coincide* (一致).

A polygon is any 2-dimensional shape on a map, i. e. *parcel of land* (地块), farmers field, lakes, or even buildings. They can be polygons on a large-scale map. Polygons show the *perimeter* (周长) and the area of the object, which can be found in the database. Particular spatial properties associated with area entities are: ① Area extent—for instance, the size of lakes; ②Perimeter length—for instance, the extent of a *shoreline* (海岸线); ③ Overlapping—for instance, *areas of circulation* (发行区, 流通区) of different newspapers, or Non-overlapping-for instance, *school districts* (学区).

(4) Volume

A volume object has length, breadth and depth, and hence is of 3 dimensions(体对象具有长度、宽度和深度，因此被称为是三维的). It is used to represent natural objects such as river *basins* (流域), or artificial phenomena such as the population potential of shopping centers or the density of residential populations(Figure 4-2).

Figure 4-2 Volume Objects Used to Present Building Blocks of Northwest A&F University

(5) Surface

A surface object is a geographical phenomenon represented as a set of continuous data, such as elevation or air temperature over an area. Surfaces can be represented by models built from regularly or irregularly spaced sample points on the surface. A geographical surface refers to, the surface of a volume object, therefore it is a kind of volume object but its depth is actually the spot height of the surface. Sometimes, it is also called 2. 5 D surface, used to represent natural or statistical surface objects(Figure 4-3).

(6) Time

Time is a dimension independent of events, in which events occur in sequence from thepast

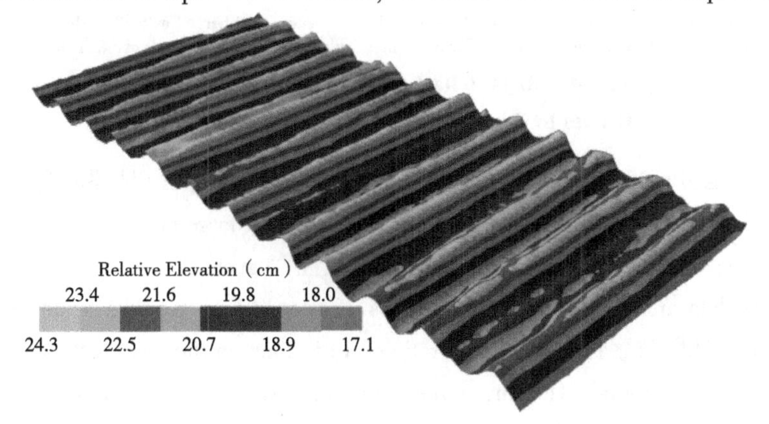

Figure 4-3 A Surface Object That Represents the 3D Surface of *Contour Tillage*(等高耕作) Slope

through the present into the future, and also the measure of durations of events and the intervals between them. Time is usually considered to be the fourth dimension of spatial objects, although GIS is currently incapable of dealing with it properly.

It is no doubt that the classification of geographical phenomena into spatial object types is dependent upon scale. For example, Beijing is represented as a 1D point on a small-scale map of the world, while a 2D area on a large-scale map of China.

4.1.3 Meaning of Scale

Scale is an essential characteristic of the spatial information. It means the size of space or the length of time as geographic information expressed. Phenomena at various scales will likely appear different and the conclusions obtained from the data can also vary(尺度是地理信息所表达的空间大小或时间长短。同一地理现象在不同尺度上的表现是不同的，而且从数据推出的结论也是不同的).

Many geographers imply a scale in the sense of *spatial resolution* (空间分辨率, the minimum distance between two adjacent features or the minimum size of a feature), or the *level of detail* (LOD, 细节度), as shown in Figure 4-4.

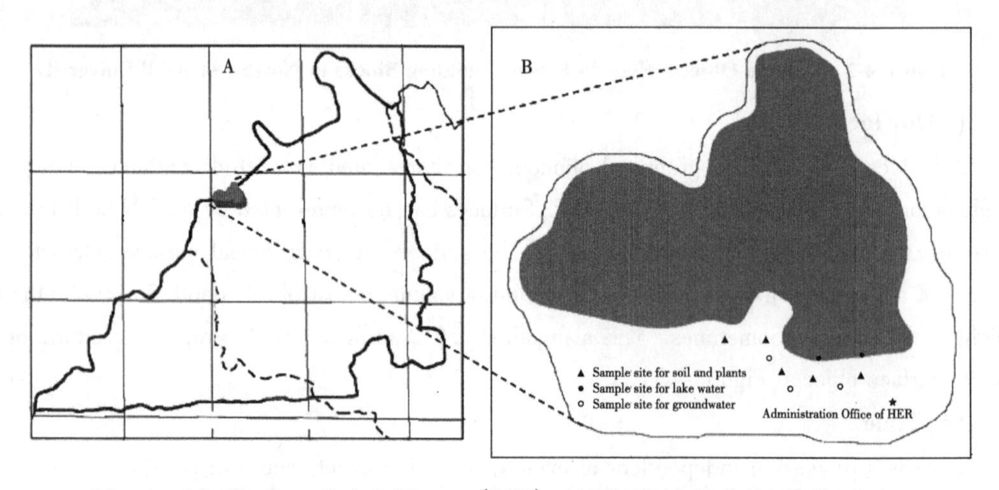

A.Location of the Hongjiannao Ecological Reserve (HER) within the Shenfu Coalfield in Medium Scale;
B. Location of Sampling Sites within HER in Finer Scale, and Include Records of Small Objects.

Figure 4-4 Maps Reflect How Spatial Resolution (Scale) Relates to the Data of The Region and Field Area

Scale is also used to imply the *geographic extent* or *scope of a project* (地理范围或项目范围). In the same spatial reference system, a large-scale project covers a large area, and a small-scale project covers a small area with the *affiliated properties* (附属属性), such as the population involved.

Geographic data are often obtained from maps. Cartographers consider scale as *a map's representative fraction* (地图数字比例尺). The ratio of a distance on the map to the *corresponding* (相应的) distance on the ground. Herein, larger scale corresponding to a large representative fraction covers a small (or small-scale) area with detailed geographic information or high spatial resolution (Table 4-2).

Table 4-2 Scale and Resolution for Some Common Map Scales

Scale		Contour Interval		Resolution of DEM (digital elevation model)
Larger Scale	1 : 500	Flat land (平地)	0. 5/1. 0 m	25 cm
		Hilly land (丘陵地)	1. 0 m	
		Upland (山地)	1. 0 m	
		Alpine (高山地)	1. 0 m	
	1 : 1000	Flat land	1. 0 m	50 cm
		Hilly land	1. 0 m	
		Upland	1. 0 m	
		Alpine	2. 0 m	
	1 : 2000	Flat land	1. 0 m	1 m
		Hilly land	1. 0 m	
		Upland	2. 0/2. 5 m	
		Alpine	2. 0/2. 5 m	
	1 : 5000	Flat land	1. 0 m	2. 5 m
		Hilly land	2. 5 m	
		Upland	5. 0 m	
		Alpine	5. 0 m	
Medium Scale	1 : 10 000	Flat land	1. 0 m	5 m
		Hilly land	2. 5 m	
		Upland	5. 0 m	
		Alpine	10. 0 m	
	1 : 25 000	Flat land	5. 0/2. 5 m	12. 5 m
		Hilly land	5. 0 m	
		Upland	10. 0 m	
		Alpine	10. 0 m	
	1 : 50 000	Flat land	10. 0/5. 0 m	25 m
		Hilly land	10. 0 m	
		Upland	20 m	
		Alpine	20 m	
	1 : 100 000	Flat land	20. 0/10. 0 m	50 m
		Hilly land	20. 0 m	
		Upland	40. 0 m	
		Alpine	40. 0 m	

(续)

Scale		Contour Interval		Resolution of DEM (digital elevation model)
Small Scale	1 : 250 000	Flat land	50. 0 m	125 m
		Hilly land	50. 0 m	
		Upland	100. 0 m	
		Alpine	100. 0 m	
	1 : 500 000	Flat land	100. 0 m	250 m
		Hilly land	100. 0 m	
		Upland	200. 0 m	
		Alpine	200. 0 m	
	1 : 1 000 000	Elevation≤2000 m	200. 0 m	500 m
		Elevation>2000 m	250. 0 m	

4. 2 Spatial Autocorrelation

4. 2. 1 Definition

Spatial objects (or spatial variables) of proximal locations usually appear a characteristic of co-variation within the geographic space (地理空间内相邻的空间对象/变量常常表现出共变的特征). That is to say, when one changes, the others within a certain distance nearby also change. This change can either be in the same direction, which is a *positive autocorrelation* (正自相关), or in the opposite direction, which is a *negative autocorrelation* (负自相关). For example, the collision between the *India plate* and the *Eurasian plate* uplifted the surface to form the *Himalayas Mountains* (印度板块和欧亚板块碰撞使地表隆起而形成喜马拉雅山脉), while the collision between the Eurasian plate and the Pacific plate made the land *subsidence* (下沉) to form the *Mariana Trench* (马里亚纳海沟). This property of co-variation between these two or more spatial objects (or spatial variables) is called *spatial dependency*(空间依赖性), while the degree of relationship between them is called spatial autocorrelation. Spatial dependency leads to the spatial autocorrelation.

Spatial autocorrelation is a kind of measurement on how near and distant things are interrelated (空间自相关是彼此相关的事物之间远近程度的度量). It is a description of the degree of similarities both in the locations of spatial objects and their attributes (空间自相关既是空间对象位置上也是属性上相似性程度的描述). If similarities exist in locations and so do they in attributes, then the pattern as a whole exhibits positive spatial autocorrelation; when features which are close together in space tend to be more dissimilar in attributes than features which are further apart, it is said to be negative spatial autocorrelation (Figure 4-5); when attributes are inde-

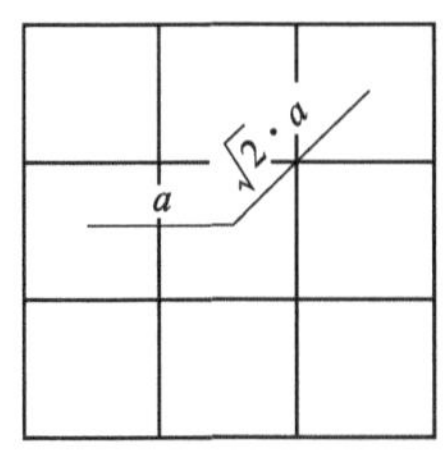

A. Positive Autocorrelation

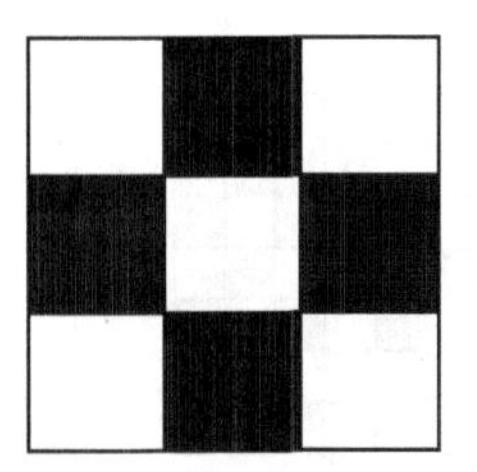

B. Negative Autocorrelation

Figure 4-5 Example of Spatial Autocorrelation (length of $\sqrt{2} \cdot a > a$)

pendent of location, it is said to be zero spatial autocorrelation, which means geographically random phenomena and *chaotic* (杂乱无章的) landscapes, and it is seldom in nature.

Similarly, temporal autocorrelation concerns the relationship between consecutive events in time. For example, someone's historical change with his/her growth. Thereby, spatio-temporal autocorrelation is the correlation of a spatial object or variable with itself over space and time (一个类似的概念，时间自相关是指时间上连续发生的事件之间的关系。因此，时空自相关是指某一个空间对象或变量在时间和空间上自身的相关性).

4.2.2 Measurement

A simple illustration of spatial autocorrelation is shown in Figure 4-6 (Longley, 2005). The features are 64 square grid arranged in the form of a *chessboard* (棋盘), of which the attributes are the two colors-black and white. Each of these five illustrations contains the same set of attributes—32 white cells and 32 black cells, but the spatial arrangements are really dissimilar. Figure 4-6A, the familiar chess board, illustrates extreme negative autocorrelation between neighboring cells [using the *Rook's or Castle's move* (国际象棋中的车或王车易位) to define neighbor]. Figure 4-6E shows the extreme positive autocorrelation, when black and white cells cluster together into homogeneous regions. The other examples in turn show independent arrangements which are intermediate at intervals along a scale of autocorrelation.

4.2.3 Autocorrelation Statistics

The autocorrelation coefficients for interval and ordinal data are mainly Moran's I, Geary's C, Getis-Ord Gi and Gi^* statistics.

Moran's I statistic gives a formal indication of the degree of linear association between a vector of observed values and a weighted average of its neighboring values. For a spatial proximity matrix W (that is, a matrix where weights reflect geographic proximity) the spatial correlation in attribute values y_i (Moran's I) is:

$$\text{Moran's } I = \frac{n \sum_{i=1}^{n} \sum_{j=1}^{n} w_{ij}(y_i - \bar{y})}{\sum_{i=1}^{n} (y_i - \bar{y})^2 \left(\sum_{i \neq j} \sum w_{ij} \right)} \tag{4-1}$$

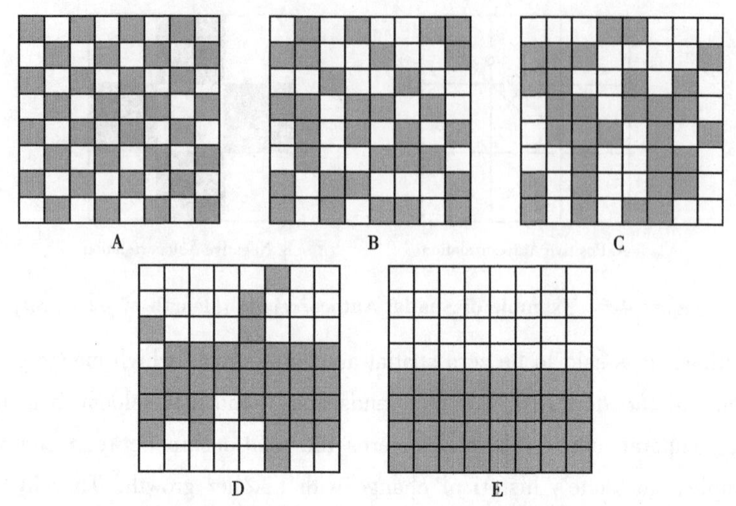

A. Extreme Negative Autocorrelation; B. Dispersed Arrangement; C. Spatial Independence (zero autocorrelation); D. Spatial Clustering; E. Extreme Positive Autocorrelation.

Figure 4-6 Simple Illustration of Spatial Autocorrelation

Where, n is the number of polygons in the study area; w_{ij} is the values of the spatial proximity matrix; y_i is the attribute under investigation; $\bar{y}$ is the mean of the attribute under investigation. Moran's I has a theoretical mean of $[-1/(n-1)]$. Thus the expected value of Moran's I is negative and is a function of the sample size n, with this expected value tending towards zero as the sample size increases. A Moran's I coefficient larger than its expected value indicates positive spatial autocorrelation: spatial clustering of similar values (either high or low). Values of Moran's I less than 0 indicate negative spatial autocorrelation: spatial clustering of dissimilar values (either high or low). Perfect negative spatial autocorrelation is represented by a checkerboard pattern.

For a spatial proximity matrix W, Geary's C is given by:

$$\text{Moran's } C = \frac{(n-1)\sum_{i=1}^{n}\sum_{j=1}^{n} w_{ij}(y_i - y_j)}{2\sum_{i=1}^{n}(y_i - \bar{y})^2(\sum\sum_{i \neq j} w_{ij})} \tag{4-2}$$

Where, n is the number of polygons in the study area; w_{ij} is the values of the spatial proximity matrix; y_i is the attribute under *investigation* (调查研究); $\bar{y}$: the mean of the attribute under investigation. The mean of the Geary statistic, under the null *hypothesis* (假设) is 1. Geary's C is never negative. Low values of Geary's C (between 0 and 1) indicate positive spatial correlation.

A different approach to quantifying spatial autocorrelation has been proposed by Getis and Ord with the Gi and Gi^* statistics. These indicate the extent to which a location is surrounded by a cluster of high or low values for the variable under consideration. They can only be computed for positive variables. The significance of the Gi statistic is assessed by a standardized Z-value. A positive and significant Z-value for the Gi statistic indicates spatial clustering of high values; while a negative and significant Z-value indicates spatial clustering of low values. Note that this interpretation is different from that of Moran's *I* statistic where spatial clustering of like values—either high or low are both indicated by positive autocorrelation.

4.3 Self-similarity & Spatial Sampling

4.3.1 Self-similarity

A coarse-scale representation ofattributes in nine squares (Figure 4-7A) is a pattern of negative spatial autocorrelation. However, when the pattern is self-replicated at finer scales (Figure 4-7B), it reveals that the smallest blue cells replicate the pattern of the whole area in a *recursive* (递归的，循环的) manner. The overall pattern is said to exhibit the property of *self-similarity*.

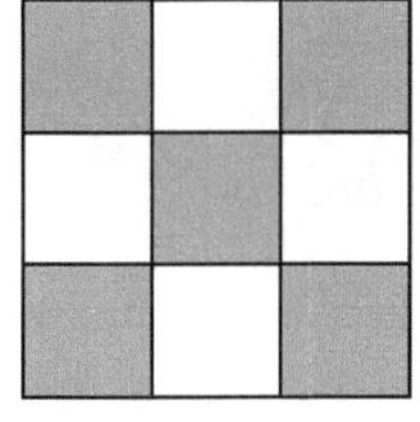

A. A Coarse-scale

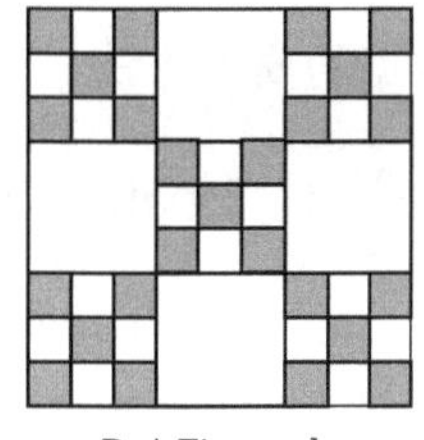

B. A Fine-scale

Figure 4-7 Representation in Self-similarity

Self-similarity is the unifying concept for the theories of Fractals and Chaos. It means the enlarged part of an object shares the same or similar features with the whole, that is, no matter how to observe an object at many different scales on a dimension [the dimension can be space (length, width) or time], there is always a finer structure and the structure is always similar. [自相似是分形和混沌理论的统一概念。自相似性是指物体局部结构放大与整体完全或大致相似的特征，即在一个维度上(空间维度如长度和宽度或时间维度]无论怎样变换尺度来观察一个物体，总是存在更精细的结构并且其结构总是相似的)(Figure 4-8).

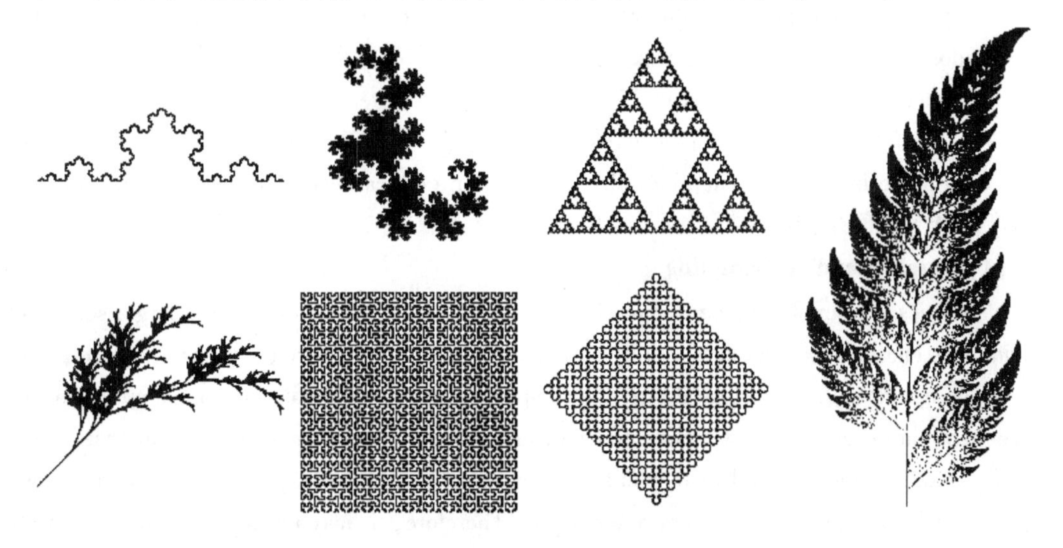

Figure 4-8 Examples of Self-similarity

Self-similar structure exists in natural systems as well as social systems (自相似结构不仅存在于自然系统，也存在于社会系统中). For example, a *rock* (岩石) may resemble the physical form of the mountain; the opinion of *Deputy to the People's Congress* (人大代表) may resemble that of the society.

In spatial data analysis, an understanding of spatial autocorrelation or self-similarity has an important influence on the way in which we abstract and collect data, and how we draw *inferences* (推理结果) between events and occurrences. Firstly, it helps us to generalize observations based on samples in order to build a spatial data set. Secondly, the presence of spatial autocorrelation *violates some of the key assumptions* (改变了一些关键的假设) of many of the conventional statistical techniques used to quantify the relationship between two or more variables. Therefore, sampling should be the typical way to obtain geographical data (因此，采样应是一种获得地理数据的典型方法).

4.3.2 Spatial Sampling

Spatialsampling involves determining a limited number of locations in geographic space for faithfully measuring phenomena that are subject to spatial dependency and heterogeneity. Spatial dependency suggests that since one location can predict the value of another location, we do not need observations in both places. However, spatial heterogeneity suggests that this relation can change across space. Therefore, we cannot fully trust a degree of dependency observed even in a small space(空间采样涉及在地理空间内具有空间依赖和异质性的有限数量的位置点的确定。空间依赖性表明，通过一个位置可以预测另一个位置的值，因此，我们就不需要在这两个地方都进行观测。但是空间异质性则表明，这种依赖关系在空间中会发生改变，所以，我们不能完全相信所观测到的数据之间的依赖性程度，即使是在一个面积小的区域中所进行的观测).

The quality of geographical data obtained can only be corresponding to that of the sampling scheme adopted(所获得地理数据的好坏仅与所用采样方案的好坏相一致). Basic spatial sampling schemes can be applied at multiple levels in a designated spatial *hierarchy* (层次) (Figure 4-9). It is also possible to exploit *ancillary data* (辅助数据), for example, using property values as a guide in a spatial sampling scheme to measure educational attainment and income (利用房产值作为空间采样的指标来衡量受教育的程度和收入).

(1) Simple Random Sampling

Instatistics, a simple random sample is a subset of individuals chosen from a larger set. Each individual is chosen randomly and entirely by chance, so that each individual has the same probability of being chosen at any stage during the sampling process. A simple random sampling is simple enough, however, areas of higher variation are not necessarily sampled more intense in this kind of sampling schema, and more points are collected than necessary in uniform areas, fewer points are collected than needed in variable areas. Therefore, it may not be as accurate as stratified sampling or as cheap as cluster sampling.

(2) Systematic Sampling

Systematic sampling relies on arranging the target observations according to some ordering scheme and then selecting elements at regular or irregular intervals through that ordered list. In

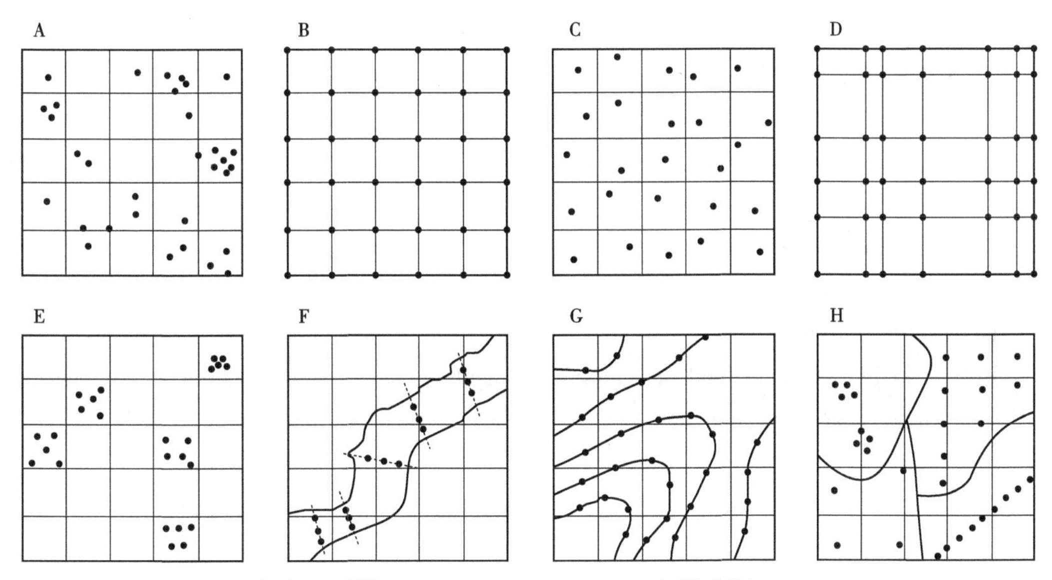

A. Simple Random Sampling（简单随机采样）; B. Regular Systematic Sampling（系统采样）; C. Systematic Sampling with Local Random Allocation（系统内随机采样）; D. Systematic Sampling with Random Variation in Grid Spacing（随机格网系统采样）; E. Cluster Sampling（聚类采样）; F. Line-intercept Sampling（横断面采样）; G. Isoline Sampling（等值线采样）; H. Stratified Sampling（分层采样 ）.

Figure 4-9 Spatial Sampling Schema

systematic sampling schema, sample points are spaced uniformly at fixed x and y intervals along parallel lines. x and y axes are not required to align with the northing and easting grid directions. It is easy in planning and description with little *subjective judgement*（主观判断）. However, this kind of sampling technology reduces statistical efficiency (equal sampling intensity in all areas and preferences cannot be addressed), and neglects the accessibility to the sample points and travel costs. Also, it may cause or rely on potentially *biased estimations*（偏差估计, inaccurate if there are patterns in the data that repeat at the same intervals), and oversampling of overproportioned areas can result in interpolation error of other locations.

(3) Cluster Sampling

Sometimes it is more cost-effective to select respondents in groups (clusters). Sampling is often clustered by geography or by time periods. Cluster sampling generally increases the variability of sample estimates of simple random sampling, depending on how the clusters differ between themselves, as compared with the within-cluster variation. If the variation between clusters is great relative to the variation within clusters, cluster sampling can result in inaccurate estimates. For this reason, cluster sampling requires a larger sample to achieve the same level of accuracy.

In clustering sampling schema, grouping of sample points is based on *defined criteria*（定义标准）: Clusters around cluster centers such that distances between points within one cluster are smaller than the distances between cluster centers; centers located randomly or systematically and sample points of each cluster may also be placed randomly or systematically around the center. This kind of technology is important for natural resource surveys due to a reduced travel time (cluster points close together), however, the variation of sampling sites are not considered explicitly.

(4) Line-intercept Sampling

Line-intercept sampling is a method of sampling elements in a region whereby an element is sampled if a chosen line segment, called a "transect", intersects the element. It is used to estimate the cover of defined classes in the landscape, for example forest or forest types. Line samples are randomly placed over the area of interest and on each sample line it is observed which proportion comes to lie in the target area class.

(5) Isoline Sampling

Isoline sampling is a method of sampling elements along an isoline, such as contour line, isotherms line, isobars line, etc.

(6) Stratified Sampling

In the stratified sampling schema, the whole surveying region can be divided into different non-overlapping blocks (strata), and one or more generalized random sample(s) can be taken around these blocks. A stratified sampling provides greater accuracy than a simple random sampling of the same size and may be less expensive because a smaller sample often provides greater precision.

4.4 Spatial Interpolation

In selectively abstracting or sampling, only part of reality is to hold within a representation. In order to represent unsampled part or the whole, we require different ways of *interpolating* (插值) to fill in the gaps between these sampling points, as well as different sample designs. Spatial interpolation is the procedure of estimating the value of properties at unsampled sites within the area covered by existing observations (空间插值是在观测区范围内通过样点的观测值来估计非采样点属性值的一个过程). In almost all cases the property must be interval or ratio scaled.

Spatial interpolation *converts* (转换) data from point observations to contiguous fields. The *rationale* (基本原理) behind spatial interpolation is that, on average, values of the attribute are more likely to be similar at points close together than at those further apart (*Tobler's Law of Geography*). This requires understanding of the *attenuating effect* (衰减效应) of distance between the sample observations, and thus of the nature of geographic data. That is to say, we need to make an informed judgment about an appropriate interpolation function and how to weight adjacent observations. The function that fills the gaps is known as *interpolation function* (插值函数).

The spatial interpolation is a very important feature of many GISs, which can be used to provide contours for displaying data graphically, to calculate some property of the surface at a given point or to change the unit of comparison when using different data structures in different layers. This technology frequently is used as an aid in the spatial decision making process both in physical and human geography as well as interrelated disciplines, such as *mineral prospecting* (矿物勘探), *hydrocarbon exploration* (油气勘探), *meteorology and climatology* (气象学与气候学).

4.4.1 Attenuating Effect of Distance

According to theliteral interpretation of *Tobler's First Law of Geography*, the *attenuation* (衰减) of a pattern or process with distance is often referred to distance decay, which describes the effect of distance on cultural or spatial interactions.

Distance decay is graphically represented by a curving line that *swoops concavely downward* (向下俯冲) as distance along the x-axis increases, and Figure 4-10 illustrates several *hypothetical* (假定的) types.

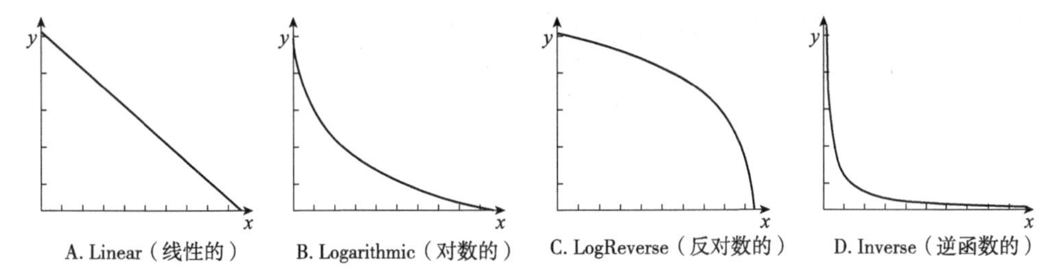

Figure 4-10 Attenuating Effect of Distance

4.4.2 Classification of Interpolation Procedures

There are several different ways to classify spatial interpolation procedures.

(1) Point Interpolation/Areal Interpolation

Point interpolation: a number of points whose locations and values are known, determine the values of other points at predetermined locations.

Areal interpolation: a set of data mapped on one set of source zones determine the values of the data for a different set of target zones.

(2) Global(全局的)/Local(局部的) Interpolation

Global interpolation determines a single function which is mapped across the whole region. Its algorithms tend to produce smoother surfaces with less abrupt changes.

Local interpolation applies an algorithm repeatedly to a small portion of the total set of points. Some local interpolators may be extended to include a large proportion of the data points in set, thus making them global.

The distinction between global and local interpolators is thus a continuum but not a *dichotomy* (二分法). This has led to some confusion and controversy in the literature.

(3) Exact(精确的)/Approximate(近似的) Interpolators

Exact interpolators honor the data points upon which the interpolation is based. The final surface passes through all points whose values are known. Approximate interpolators, *B-splines* (B样条) and *Kriging* (克里格) methods all honor the given data points.

Approximate interpolators are used when there is some uncertainty about the given surface values. This utilizes the belief that in many data sets there are global trends, which vary slowly, overlain by local fluctuations, which vary rapidly and produce uncertainty (error) in the recorded values. The effect of smoothing will therefore be to reduce the effects of error on the resulting surface.

(4) Stochastic(随机的)/Deterministic(确定的) Interpolators

Stochastic methods incorporate the concept of randomness. The interpolated surface is conceptualized as one of many that might have been observed, all of which could have produced the known data points. Stochastic interpolators include trend surface analysis, *Fourier analysis* (傅里叶分析) and Kriging. Procedures such as trend surface analysis allow the statistical significance of the surface and uncertainty of the predicted values to be calculated.

Deterministic methods do not use probability theory.

(5) Gradual(渐进的)/Abrupt(突变的) Interpolators

A typical example of a gradual interpolator is the distance weighted moving average. Usually it produces an interpolated surface with gradual changes. However, if the number of points used in the moving average is reduced to a small number, or even one, there would be abrupt changes in the surface.

It may be necessary to include barriers in the abrupt interpolation process. Semipermeable barriers (e. g. weather fronts) will produce quickly changing but continuous values. Impermeable barriers (e. g. geologic faults) will produce abrupt changes.

4. 4. 3 Popular Methods of Interpolation

Mostly, data for interpolation come from sample attributes at few points. These measurements are known as 'hard data'. Information on physical processes or phenomenon that causes the pattern is called 'soft data'(Figure 4-11).

There are various methods for conducting the interpolation.

(1) Inverse Distance Weighting

Inverse distance weighting (IDW, 反距离权重) is a common and simple spatial interpolation method. It combines the weighted average distance between the interpolation point and the sample point, that is, sample point from the more closer interpolation points is given the greater weight (它以插值点与样本点间的距离为权重进行加权平均，即离插值点越近的样本赋予的权重越大). This method is simple, intuitive, and efficient. The interpration effect is better when sample points are uniformly distributed. The interpolation result is between the maximum and minimum values of the sample data. But the disadvantage ot this method is that the resulting data is susceptible to the impact of extreme value (该方法简单易行，直观且效率高。在已知采样点分布均匀的情况下插值效果好，插值结果在用于插值的样点数据的最大值和最小值之间。缺点是易受极值的影响).

(2) Spline

Using a mathematical polynomial fitting function, spline interpolation generates a smooth interpolation curve by controlling the estimated variance of some limited point values based on the feature of some special nodes. This method is applicable to a gradual change in the surface, such

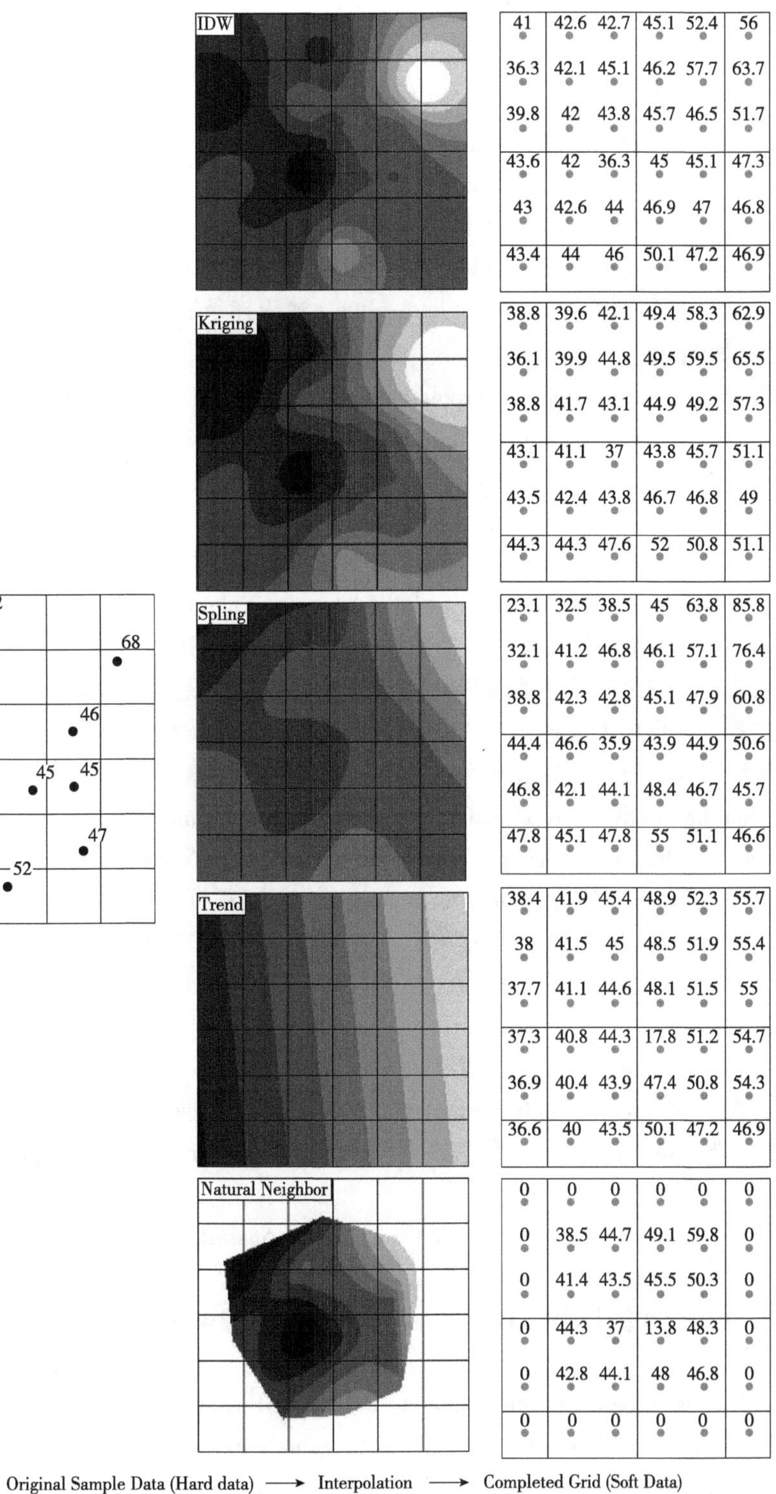

Figure 4-11 Example of Interpolation

as temperature, elevation, groundwater height or pollution concentration. This method is easy to operate with little computation. But it is difficult to estimate the error, and it doesn't work well with scarce number of sample points (样条插值法是利用一种数学多项式拟合函数，对一些限定的点值，通过控制估计方差，利用一些特征节点，产生平滑的插值曲线。这种方法适用于逐渐变化的曲面，如温度、高程、地下水位高度或污染浓度等。该方法优点是易操作，计算量不大，缺点是难以对误差进行估计，采样点稀少时效果不好).

Spline interpolation is divided into spline with tension (张力样条插值法), regularized spline (规则样条插值法), thin-plate spline (薄板样条插值法).

(3) Kriging

The basis of the Kriging interpolation is the rate at which the variance between points changes over space. This is expressed in the variogram which shows how the average difference between values at points changes with distance between points. Kriging takes into account the location relationship of the observation points and the estimated points, and also considers the relative location relationship between observation points. It is better than inverse distance weighting interpolation when the sample points are scarce as the Kriging interpolation method for spatial data often achieve desired results (克里金插值方法是建立在空间中样点间的方差变化速率的基础上，用变异函数来表示点值之间的平均差异随间距的变化情况。该方法考虑了观测的点和被估计点的位置关系，并且也考虑各观测点之间的相对位置关系，在样点稀少时插值效果比反距离权重等方法要好，所以利用克里金方法进行空间数据插值往往取得理想的效果).

According to different objectives, a variety of kriging methods are developed, such as: Simple-Kriging, Ordinary-Kriging, Universal-Kriging, Log-Normal Kriging, Cokriging, Pseudo-Kriging, Indicator-Kriging, and Disjunctive-Kriging (根据应用目标不同，发展了多种克里格方法如：简单克里格、普通克里格、泛克里格、对数正态克里格、协同克里格、拟协克里格、指示克里格、离析克里格).

Besides these 3 commonly used interplation methods, there are several interplation methods as classification models (分类模型), trend surface analysis (趋势面分析), Fourier series (傅里叶级数), discrete smooth interpolation(DSI-离散平滑插值), least squares method (最小二乘法), Thiessen polygons (泰森多边形), pycnophylactic methods (Pycnophylactic 插值), and so on.

More detailed info. please check the course of *Spatial Analysis*.

4.5 Uncertainty of Geographical Data

GIS allows geographic phenomena to be depicted using vector and raster data. These data can then be used to analyze the phenomena to gain better understanding as to why or how they occurred. When working with a geographic phenomenon, there is always a degree of uncertainty. The concept of "uncertainty" is defined as the difference between a real geographic phenomenon and the user's understanding of the geographic phenomenon. A user may view certain relationships occurring and often draw cause-affect conclusions (当用户看到特定的关系发生时常常会由此而推演出一些结论). The factual cause-effect relationship may be different and the perceived geographic pattern may have been misunderstood.

Uncertainty in geographic representation arises inevitably because almost all representations of the world are incomplete (由于几乎所有的对世界的表达都是不完整的，所以地理表达的不确定性就不可避免地出现了). Uncertainty arises from the way that GIS users conceive of the world, they measure and represent it, and they analyze their representations of it, such as the length of coast line problem.

4.5.1 The Length of Coast Line Problem

Length of the coastline is infinite—The coastline length depends on the unit of measurement. For example, with 1 km as measurement unit, we can obtain an approximate result in which the twists length shorter than 1km will be ignored; with 1m as measurement unit, the former ignored tortuous length (less than 1km and large than 1m) will be measured, therefore the length of the coastline change increases. When the measurement unit becomes smaller and smaller, the resulting length will infinitely increase. The length of the coastline increases with the increasing level of details and scale (海岸线的长度随着细节度与尺度的增加而增加). In the end, the length will tend toward a *determined value* (确定值), this *limit value* (极限值) may be the length of the coastline.

Naturally, the coastline is very irregular, such as the coastline in the *tidal* (潮汐). On the other hand, we just linearize it when surveying, therefore the coastline is extremely unsmooth. Thus we can safely draw a conclusion that the precise length of the coastline is indeterminate and scale-dependent.

4.5.2 Uncertainty of Geographical Phenomena

Uncertainty may be defined as a measure of the difference between the data and the meaning attached to the data.

(1) U1: Uncertainty in Geographical Concepts

Many common geographical concepts are vague, such as "near" "north" "mountain" and "developing world". Therefore, in many cases there are no natural units of geographical analysis. What is the exact distance from a specific object? What is the natural unit of measurement for a soil profile? … It is really difficult to answer these questions accurately.

Many linguistic terms used to convey geographic information are also inherently ambiguous (用于传达地理信息的很多术语本身也具有歧义), such as "*weight* (权重)" "*unit area* (单位面积)" "per capita(人均)". Many objects are assigned different labels by different national or cultural groups, and such groups perceive space differently. Geographic *prepositions* (介词) like *across*, *over*, and *in* do not have simple correspondences with terms in other languages. Object names and the topological relations between them may thus be inherently *ambiguous*.

(2) U2: Uncertainty in Measurement

When measuring or calculating some quantity from geographic data, we generally assume that some exact or "true values" exist based on how we define what is being measured (or calculated). The results are usually reported as a range of values that we expect this "true value" to fall within. The most common way to show the range of values is: measurement = best estimate ± uncertainty. All measurements have a degree of uncertainty regardless of precision and accuracy. This is caused by two factors: the limitation of the measuring instrument (systematic error) and the skill of the experimenter making the measurements (random error).

Uncertainty in measurement exists in capturing, digitizing, editing geographic data, and data integration and data sharing (量测中的不确定性存在于地理数据的获取、数字化、编辑及数据整合和数据共享过程中).

(3) U3: Uncertainty in Analysis

Undoubtedly, uncertainty in data will lead to uncertainty in the results of analysis (无疑，数据的不确定性将导致分析结果的不确定性). For example, when two measurements are added (or subtracted, multiplied, divided), the (percentage) uncertainty of the results is always equal to the sum of the (percentage) uncertainties of the two measurements. As a result, the analysis may be inappropriately influenced by aggregate data (作为一个结果，分析可能收到汇总数据的不当影响). Inappropriate inference from aggregate data about the characteristics of individuals is termed the ecological fallacy (由单个数据特征汇总而得到的数据所做的不当推论被称为生态谬误). An ecological fallacy (or ecological inference fallacy) is a logical fallacy: the correlation between individual variables is deduced from the correlation of the variables collected from the group to which those individuals belong.

Vocabulary

autocorrelation[ˈɔ: təu, kɔrəˈleiʃən] *n.* 自相关
exploration[ekspləˈreɪʃ(ə)n] *n.* 勘探
heterogeneity [ˌhetərəˈdʒəˈniːəti] *n.* 多相性；异质性
hydrocarbon[ˌhaɪdrəˈkɑːrbən] *n.* 碳氢化合物
hypothetical[ˌhaɪpəˈθetɪkl] *adj.* 假设的；假定的
indicator[ˈɪndɪkeɪtə] *n.* 指示器
interpolation[ɪnˌtɜːpəˈleɪʃn] *n.* 插话；添加内容；[数]插值
interrelated[ˌintəriˈleitid] *adj.* 相互联系的
landscape[ˈlæn(d)skeɪp] *n.* 风景
linguistic[lɪŋˈgwɪstɪk] *adj.* 语言的
mineral prospecting 矿物勘探
oscillate[ˈɒsɪleɪt] *v.* 振荡
parameter [pəˈræmɪtə] *n.* 参数
pragmatic[prægˈmætɪk] *adj.* 实用的

rational[ˈræʃ(ə)n(ə)l] *n.* 有理数
inherently[inˈhiərəntli] *adv.* 内在的
replicate[ˈreplɪkeɪt] *v.* 折叠 *adj.* 复制的
spatio-temporal 时空的
self-similarity 自相似

Questions for Further Study

1. What is thedefinition of the term *scale*, and list its different uses.

2. How many aspects do the nature of geographic data cover?

3. Design different sampling schema you may adopt in your studies, and compare their advantagesand disadvantages.

4. Based on the understanding of the concept of spatial interpolation, try to summerize different spatial interpolation methods.

5. What is the meaning of uncertainty of geographic data?

6. Explain how ecological fallacy arises and why it can lead to false conclusions.

Chapter 5

GIS Data Modeling

The real world is infinitely complicated. We need to adopt all kinds of different structuring models when we present the real world in geographical system and analyze it. We also need to simplify its inherent complexity. We perceive the real world differently and differentiate purposes for using a geo-information system, so it is necessary to choose a data model for us to present the world. A data model in GIS is a mathematical construct for representing geographic objects as data. Data models are *vitally*（非常地，极其地）important to GIS because they control the way data are stored and have a major impact on the type of analytical operations performed. Whenever you deal with geographic data, you must choose a data model for operating them. Increasingly, GIS models are used in many disciplines to provide valuable insights into problems such as evaluating locations, analyzing movement and predicting the trend.

5.1 Basic concepts

5.1.1 Model

In the most general sense, a model is something used in some way to represent something else. The model is a simplified representation of reality used to simulate a process, understand a situation, predict an outcome, or analyze a problem. A model can be viewed as a selective approximation which, by elimination of incidental detail, allows some fundamental aspects of the real world to appear or be tested（模型是对现实的简化表达，用来模拟一个过程，理解一种情况，预测一种结果或分析某一问题。模型可被看作一种具有选择性的近似表达，它通过排除偶然的细节来对真实世界的某些基本方面进行显示或测试）. A model is structured as a set of rules and procedures, including spatial modeling tools available in a GIS.

Modeling specifies data transformations which involvethe *synthesis of information*（信息综合）. The "synthesis" is the process of putting together expressions of general principles with representations of parts of the reference system so as to form a *replica*（复制品）that exhibits behavior similar to that of the reference system.

Modeling can be defined in the context of GIS as occurring whenever operations of the GIS attempt to emulate processes in the real world, at one point in time or over an extended period. Types of models include:

(1) Conceptual Model(概念模型)

Qualitative models that help highlight important connections in the real-world systems and processes. They are used as the first step in the development of more complex models—a set of concepts that describe a subject and allow reasoning about it.

(2) Logical Model(逻辑模型)

Implementation-oriented (面向应用的) representation of the reality used to describe the structure of some information domain. This consists of descriptions of (for example) tables, columns, object-oriented classes, and XML *tags* (标签).

(3) Physical Model(物理模型)

A smaller or larger physical copy of an object so that interactive demonstrations can be easily observed and manipulated. Physical models have characteristics similar to key features of more complex systems in the real world. These models can help bridge the gap between conceptual models and models of more complex real-world systems.

(4) Mathematical and Statistical Model

①Mathematical Models originate from relevant equations that determine how a system changes from one state to the next (*Differential Equations*, 微分方程) and/or how one variable depends on the value or state of other variables (*State Equations*, 状态方程). These can also be divided into either *Numerical models* (数值模型) or *Analytical models* (分析模型).

②Statistical Models include issues such as statistical characterization of numerical data, estimating the probabilistic future behavior of a system based on past behavior, *extrapolation* (外推法) or interpolation of data based on some best-fit, error estimates of observations, or spectral analysis of data or model generated output. These models are useful in helping identify patterns and underlying relationships between data sets.

5.1.2 Data Model

(1) Basic Concept

Basically, a data model in many *fields* (领域) can be considered as an abstract model or a *medium* (媒介) that documents and organizes *particular data* (特定的数据) for communication between team members from different backgrounds and with different levels of experience, and it is used to explain how data are needed, created, *stored*, *accessed* (存取), or analyzed. As a key factor of a data model used to exchange data, precision means that the *terms and rules* (术语和条款) on a data model can be interpreted only one way and are not *ambiguous* (模糊不清的). Sometimes, a data model can be referred to as a *data structure* (数据结构), especially in the context of *programming languages* (程序设计语言). Data models are often *complemented* (相辅相成) by *function models* (函数模型), especially *in the context of* (在…的情况下) enterprise models.

The core of any GIS is the data model, which is a set of constructs for describing and representing selected aspects of the real world in a computer. People interact with operational GIS in order to *undertake*（承担）tasks like making maps, querying databases, and performing spatial analysis.

The geographic reality is continuous and infinitely complex, but computers are finite and can only deal with digitized data. Therefore, it is difficult to make choices about what things are modeled in a GIS and how they are represented. Since different people use GIS for different purposes and even a single object would have different characteristics, there is no single type of *all-encompassing*(包罗万象的) GIS data model best for all circumstances.

(2) Levels of Abstraction

There are 3 kinds of data model instance（实例）except real world itself when representing the real world in a computer (Figure 5-1).

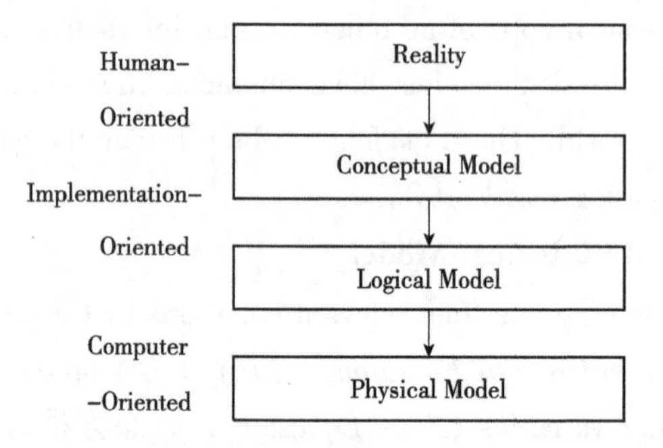

Figure 5-1 Levels of Data Models with Increasing Abstraction

Conceptual models refer to models that exist only in human mind and represent human *intentions*（意图）or *semantics*（语义）and help us know and understand the real world they represent. Therefore, the conceptual model is a *human-oriented*（以人为目标的）, often partially structured model of selected objects. Concepts are used to *convey semantics*（传达语义）in various natural languages.

Logical data models represent the abstract structure of a domain of information. They are often *diagrammatic*（图表的）in nature and typically used in *business processes*（业务流程）that seek to capture things of importance to an organization and how they relate to one another. Once *validated*（验证）and approved, the logical data model can become the basis of a *physical data model*（物理数据模型）.

A physical data model (or database design) is a representation of a data design which *takes into account*(考虑) the facilities and constraints of a given database management system. In the lifecycle of a project, it typically *derives from*（源于）a logical data model. A complete physical data model will *portray*（描绘）the actual implementation in a GIS, and often comprises all the database *artifacts*（人工制品, such as tables stored as files or databases) required to create relationships between tables or to achieve performance goals, such as indexes, constraint definitions, linking tables, partitioned tables or clusters.

5.2 GIS Data Model

In GIS, data model is a kind of conceptual model for describing and reasoning about the world expressed in a data structure (e. g. ASCII files, Excel tables...) or a geographic database. Data modeling can be defined as occurring whenever operations of the GIS attempt to *emulate* (模拟) processes in the real world, at one point in time or over an extended period. It helps us explore the basic questions behind GIS analyses, understand the spatial concepts that underlie those questions, and determine appropriate methods to employ when performing GIS analyses.

Geographical data model (a GIS database) is a model of the real world that can be used to mimic certain aspects of reality. Geographical data model is designed to mimiconly selected aspects of reality; it must represent certain entities and relationships among them; it may be represented in words, mathematical equations or maps or a GIS representing spatial relations. It can be tested and manipulated more conveniently at a faster (or slower) rate and less expensively than the condition it mimics; it is used to answer questions about what exists now, what existed at some point in the past, and what will happen in the future; it is used when it is more convenient or it is not possible to collect the information directly. e. g. It is convenient to measure road distance on a map; The height a forest will reach in 5 years' time is impossible to measure directly. Anyway, a more complex geographical data model may or may not provide "better" answer.

Traditionally, spatial data was stored and presented in maps. Maps are the original spatial database, but they are not so good for updating and analyzing. We have to tell computers how to perform most of the interpretation of maps that humans do that lead to ways to represent and model reality. The things we do *cognitively*(认知地) manually often become difficult to teach a computer to do. We must define all terms and features, quantify.

A collection of entities of the same geometric type *is referred to as*(被称为) a class or layer. It should also be noted that the term layer is quite widely used in GIS as a generic term for a specific dataset. It derived from the process of entering different types of data into a GIS from paper maps, which was undertaken one plate at a time. Grouping entities of the same geographic type together makes the storage of geographic databases more efficient. It also makes it much easier to implement rules for validating edit operations and for building relationships between entities. All of the data models use layers in some way to handle geographic entities. Therefore, a layer is a collection of geographic entities of the same geometric type (e. g. points, lines, or polygons) (Figure 5-2). Grouped layers may combine layers of different geometric types.

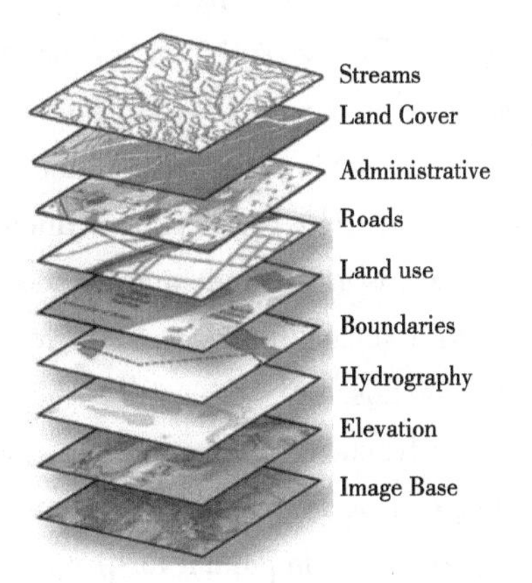

Figure 5-2 Geographic Data Model Based on Inventory of Data Layers

A GIS data model is a mathematical construct for representing geographic objects as data. For example, thevector data model represents geography as collections of points, lines, and polygons; the raster data model represents geography as *cell matrices* (单元矩阵) that store numeric values; the TIN data model represents geography as sets of contiguous, non-overlapping triangles.

5. 2. 1 The Vector Data Model

The vector data model is closely linked with the discrete object conceptual data model and uses a set of coordinates and relationships to represent the real- world objects. Geographic entities encoded in the vector data model are usually called features. Features of the same geometric type are stored in a geographic database as a feature class. the physical (database) representation stored as a feature table is preferred. GIS generally deals with two types of feature: simple (non-topological) and topological (Figure 5-3).

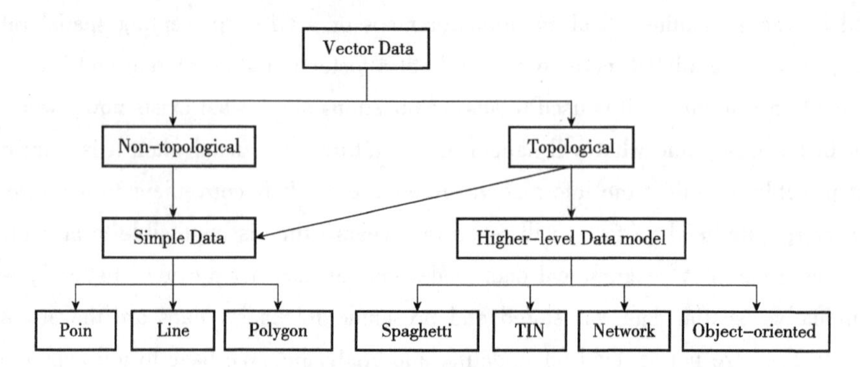

Figure 5-3 Composition of Vector Data

(1) Simple Data Model

As introduced in Chapter 3 and Chapter 4, the basic unit in vector storage is a point (X, Y coordinate). Connected points make lines. Lines may enclose a polygon. Vector storage is object-based (i. e. feature based). It tends to store geographic features as objects; i. e. a line stored *sequentially* (循序的) in the database. It will digitize (or trace) in the same way as we draw maps. It is important for objects in features to be in order; i. e. points and segments. The order of objects is not important; i. e. one line before another is not important. A discrete data model specifies the *distinct* (明显的) location, and has specific edges. Change is abrupt (hence discrete). Location is an attribute and location of the entity must be stored. *Adjacency*(邻接性) is not inherent in a vector file. A typical file says nothing about what is next to what. We need *topology*(拓扑) for this.

Advantages of simple data models are: ①They are easy to be created, stored and retrieved. They can be *rendered* (呈现) on a computer monitor more rapidly than topology-based data. This is particularly important for people who use, rather than produce, GIS data. ②They are *non-proprietary and interoperable* (非专有的、可互操作的). It means that they can be used across different software packages. In addition, the simple features belong to the non-topological structured data, and they lack more advanced data structure characteristics such as topology. Operations of spatial data analysis, like shortest-path network analysis, polygon adjacency, etc. cannot be performed without additional calculations.

Spaghetti data is a kind of vector data composed of simple lines with no topology stored and usually no attributes (Figure 5-4). In such a model, the geometry of any spatial object is independent of other objects(对于这样的模型，任一空间对象的几何形状均独立于其他对象). Spaghetti lines may cross, but no intersections are created at those crossings. Therefore, their topological relationships must be computed on demand.

Points

Point Number	Coordinates (*x*, *y*)
1	(3, 6)
2	(4, 3)
3	(5, 5)
4	(6, 2)

Polylines

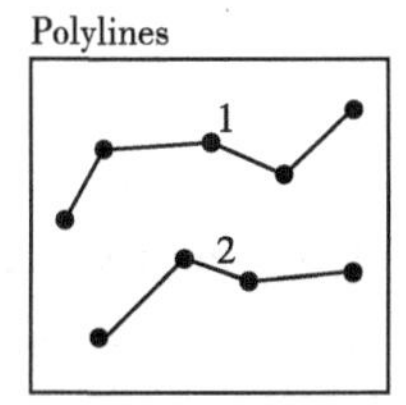

Line Number	Coordinates (*x*, *y*)
1	(1, 4) (2, 5) (4, 6) (3, 4) (5, 6)
2	(2, 1) (3, 3) (4, 2) (5, 3)

Polygons

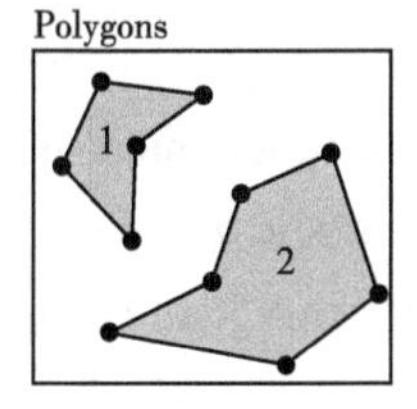

Polygon Number	Coordinates (*x*, *y*)
1	(1, 4) (2, 7) (5, 6) (3, 5) (3, 3)
2	(3, 1) (5, 3) (6, 4) (7, 6) (8, 3) (6, 1)

Figure 5-4 Sample of Simple Data

(2) Topological Data Model

Topology is the mathematical science of geometrical relationships to validate the geometry of vector entities and for operations such as network tracing and tests of polygon adjacency. Topology in GIS refers to relationship or connectivity between spatial objects. Topological relationships include *proximity* (关联性), *connectivity* (连通性), *adjacency* (邻接性), *membership or inclusion*(包含性), and *orientation* (方向性). The explicit representation of such relationships in the spatial data model provides more knowledge, which is helpful for the evaluation of queries (在空间数据模型的显示表达关系中提供了更多的知识点，这对评估查询很有帮助).

Topological relationship is the non-metric (qualitative) property of geographic objects that *remain constant* (保持不变) when the geographic space of objects is *distorted* (扭曲变形)(Figure 5-5). For example, when a map is stretched, non-topological properties such as distance and angle change, whereas topological properties such as adjacency and containment do not. Topology

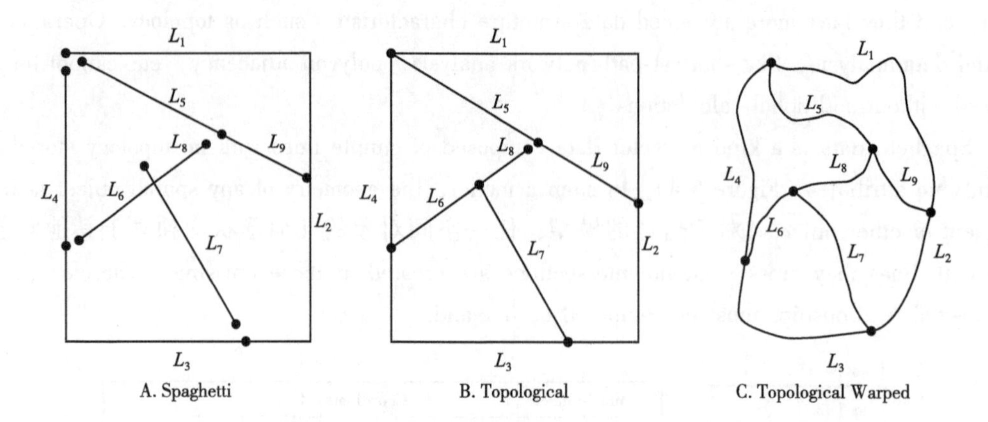

Figure 5-5 Spaghetti, Topological, and Topological-Warped Vector Data

(B and C are topologically identical because of their same connectivity and adjacency)

is important in GIS because of its role in data validation, Modeling integrated features behavior, editing, and query optimization.

(3) Network Data Model

The network data model is really a special type of topological feature model. The network model was first designed for representing networks in network-based application such as transportation services or utility management (electricity, telephone). In this model, *topological relationships*(拓扑关系) among points and polylines are stored. The set of geometric types of network is slightly more complex than in the spaghetti model.

In GIS software systems, networks are modeled as points (i. e. street intersections, fuses, switches, *water valves* (水阀), and the confluence of stream reaches in network. they usually referred to as nodes in topological models-A node is distinguished point that connects a list of arcs), and lines (i. e. streets, transmission lines, pipes, stream reaches: usually referred to as arcs— An arc is a polyline that starts at a node and ends at a node). Network topological relationships show how lines connect with each other at nodes. For the purpose of network analysis (see Chapter 10), it is also useful to display rules about how flows can move through a network.

In geo-relational implementations of the topological network feature model, the geometrical and topological information is typically held in ordinary computer files and the attributes in a linked database. The GIS software tools are responsible for creating and maintaining the topological information each time and there is a change in the feature geometry. In more modern object models, the geometry, attributes, and topology may be stored together in a DBMS, or topology may be computed on the fly. One advantage of this approach is its *intrinsic* (本质的) description of a networked topology. With the notion of connectivity, it is useful for *optimal-path* (最佳路径) search. No information on the relationships between 2D objects is stored in this model.

(4) TIN Data Model

The *triangulated irregular network* (TIN, 不规则三角网) is a commonly used data structure in GIS. It approximately represents the terrain with a set of non-overlapping triangles (用一系列

不重叠三角形近似表示地形). It is one of the standard implementation techniques for *digital terrain models* (DTM, 数字地形模型), but it can be used to represent any continuous field (Figure 5-6).

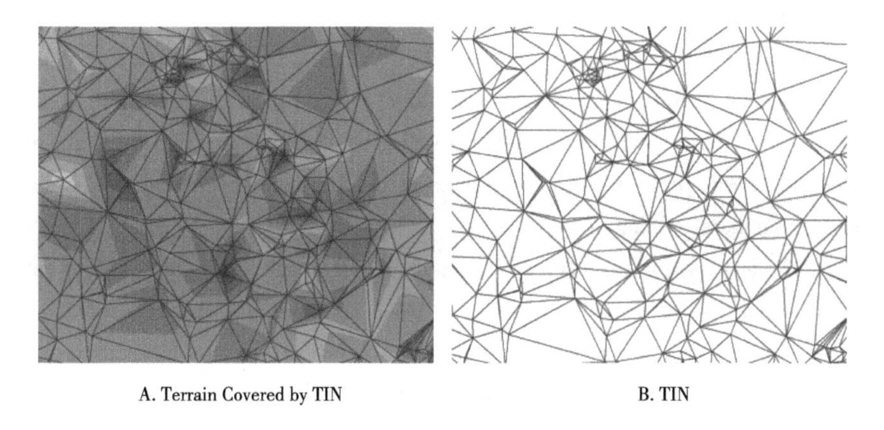

A. Terrain Covered by TIN　　B. TIN

Figure 5-6 A TIN Uses A Series of Non-Overlapping Triangles to Represent the Terrain

The principles behind a TIN are simple. Itis built from a set of known locations, for instance an elevation. The locations can be arbitrarily scattered in space (位置任意散布在空间内), and are usually not on a nice regular grid or a straight line. Any location together with its elevation value can be viewed as a point in three-dimensional space. From these 3D points, we can construct an irregular tessellation made of triangles (Figure 5-7).

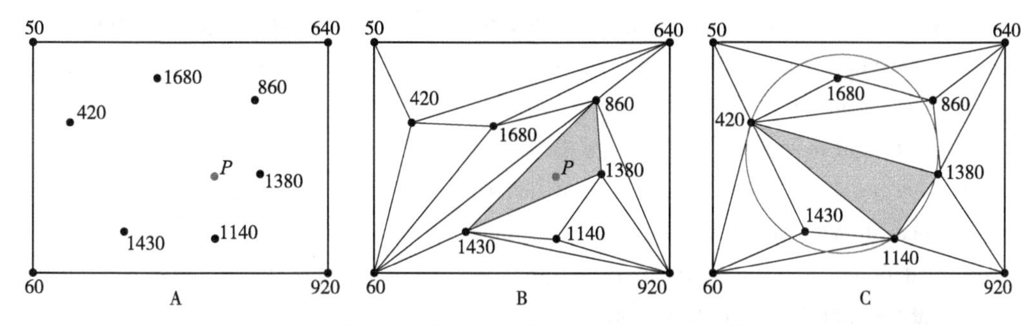

A. Input Locations and Their (elevation) Values. The Location P is an Arbitrary Location That Has No Associated Elevation Measurement and That is Only Included for Explanation Purposes; B. Triangulation with Many 'stretched' Triangles Based on the Input Locations; C. Triangulation is More Equilateral.

Figure 5-7 A TIN Construction

In the three-dimensional space three points *uniquely* (唯一地) determine a plane as long as they are not *collinear* (they must not be positioned on the same line). A plane formed by these points has a fixed *aspect* (方向) and *gradient* (坡度) and can be used to compute an *approximation* (近似值) of elevation of other locations. Since we can pick many *triples of points* (三角形的点), we can construct many such planes, and therefore we can have many elevation approximation for a single location, such as point P. So, it is wise to restrict the use of a plane to the triangular area ‘between’ the three points.

If we restrict the use of a plane to the area between its three *anchor* (固定的) points, we will obtain a *triangular tessellation* (三角网) of the complete study space. Unfortunately, there are many different tessellations for a particular input set of anchor points, as Figure 5-7 demonstrates with two of them. Some tessellations are better than others in the sense that they make smaller errors of elevation approximation. For instance, if we base our elevation computation for location P on the left- hand shaded triangle, we will get another value from the right-hand shaded triangle. The second will provide a better approximation because the average distance from P to the three triangle anchors is smaller.

A TIN clearly is a vector representation: eachanchor point has a stored geo-reference. We might also call it an irregular tessellation as the chosen triangulation provides a *tiling* (瓦片) of the entire study space. The value of cells of this tiling, however, do not have an associated stored value as typical of tessellations, but it is obtained by a simple *interpolation* (内插) of the elevation values of the three anchor points.

(5) Object-Oriented Vector Data Model

In the object-oriented approach, we model theworld by objects. Before applying the approach to a real world problem, we need to understand what an object really is.

A *class* (类) is an entity that has a well-defined role in the application domain. The organization wishes to maintain state, behavior, and identity. A class is a concept, abstraction, or thing that makes sense in an application context. A class could be a *tangible* (实在的) or visible entity (e. g. a person, place, or thing); it could be a concept or event (e. g. Department, Performance, Marriage, Registration, etc.); or it could be an artifact of the design process (e. g. User Interface, Controller, Scheduler, etc.).

An object is an instance of a class (对象是类的实例) (e. g. a particular person, place, or thing) that *encapsulates* (封装) the data and behavior we need to maintain about that object. A class of objects shares a common set of attributes and behaviors. An object is the basic atomic unit in an object data model and comprises all the properties that define the state of an object, together with the methods that define its behavior.

The state of an object *encompasses* (包含) its properties (attributes and relationships) and the values those properties have, and its behavior represents how an object acts and reacts. An object's state is determined by its attribute values and links to other objects. An object's behavior depends on its state and the operation being performed. An operation is simply an action that one object performs upon another in order to get a response. You can think of an operation as a service provided by an object (supplier) to its clients. A client sends a message to a supplier, which delivers the desired service by executing the corresponding operation.

Examples of objects include oil wells, soil bodies, *stream catchments* (支流集水量), and aircraft flight paths. In the class of oil wells, each oil well might include properties defining its state -annual production, owner name, the date of construction, and the type of geometry used for representation at a given scale (perhaps a point on a small-scale map and a polygon on a large-scale one). The oil well class could have connectivity relationships with a pipeline class that represents the pipeline used to transfer oil to a *refinery* (冶炼厂). There could also be a relationship defining the fact that each well must be located on a *drilling platform* (钻井平台). Finally, each oil well might also have methods defining the behavior or what it can do. For example, behavior might include how objects draw themselves on a computer screen, how objects can be created, deleted, and edited rules, and how oil wells snap to pipelines.

There are three key *facets* (刻面) of object-oriented data models that make them especially good for modeling geographic systems: *encapsulation* (封装), *inheritance* (继承), and *polymorphism* (多态). Encapsulation describes the fact that each object packages together a description of its state and behavior. The state of an object can be thought as its properties or attributes. Inheritance is the ability to reuse some or all of the characteristics of one object in another object. Polymorphism describes the process whereby each object has its own specific implementation for operations like drawing, creating, and deleting.

All geographic objects have some type of relations with other objects in the same class or possibly with objects in otherclasses (所有的地理对象均与同一对象类中或其他对象类中的对象具有某种关系). Some of these relationships are inherent in the class definition (e. g. some GIS removes overlapping polygons) while other *interclass* (组内的) relationships are *user-definable* (用户可定义的). Three types of relationships are commonly used in geographic object data models: topological, geographic, and general.

Generally, topological relationships are constructed into the class definition. Geographic relationships between object classes are based on geographic operators (such as coincidence, adjacency, containment, and touch) that determine the interaction between objects. General relationships areuseful to define other types of relationship between objects.

In addition to supporting relationships between objects (strictly speaking, between object classes), object data models allow several types of rules to be defined. Rules are the valuable means of maintaining database integrity during editing tasks. The most popular types of rules used in object datamodel are attribute, connectivity, relationship, and location.

Attribute rules are used to define the possible attribute values that can be entered for any object. Both range attribute rules and coded value attribute rules are widely employed. A range attribute rule defines the range of valid values that can be entered. Coded value attribute rules are used for categorical data types. *Connectivity rules* are based on the specification of valid combinations of features, *derived* (导出) from the geometry, topology, and attribute properties. *Geographic rules* define what happens to the properties of objects when an editor *splits* (分裂) or merges them. In the case of a land parcel split following the sale of part of the parcel, it is useful to define rules to determine the impact on properties like area, land use code, and owner.

(6) Vector Storage and Encoding Techniques

Vector data are stored as aseries of x, y coordinates (Figure 5-4).

①Points are singlelocations or occurrences; they are dimensionless without length, width, or area. They are stored as single x, y coordinates (a "z" value can be added), such as wells, buildings.

②Lines are a string of x, y coordinates. They (also called "arcs") have the linear feature of following a path between two points. They have length but no width or area.

③Polygons are composed of line(s) and labels. Polygons (also called area) are composed of a set of connected lines (arcs) within enclosed space . They are isolated (独立的), adjacent, or nested (嵌套). There is a homogeneity (同质性) within the area.

④Attribute data are used torecord the non-spatial characteristics of an entity. Attributes are attached to each feature through a unique numeric code. They may be stored as tables with attributes arranged in rows and columns in each vector file. Each row corresponds to an individual spatial object, and each column corresponds to an attribute.

(7) Spaghetti Data Structure and Data Encoding

The spaghetti data structure is a simple data structure that stores the data in an unstructured way. This data structure just stores the name of every object and the coordinates the object is composed of. Since the objects are not related to each other, no topological information is included and the consistency cannot be verified. Every object of the spaghetti data structure is described independently of the others. The same coordinates may appear several times, therefore large amounts of storage space is required. In addition, the advantage of this data structure is the possibility to modify every object without affecting the others. Each point, line, or polygon that represents a geographic entity is stored as a record in a file that consists of that entity's ID and a list of coordinates that define its location (or the coordinate space that it occupies) (Figure 5-8).

Characteristics of spaghetti data structure:

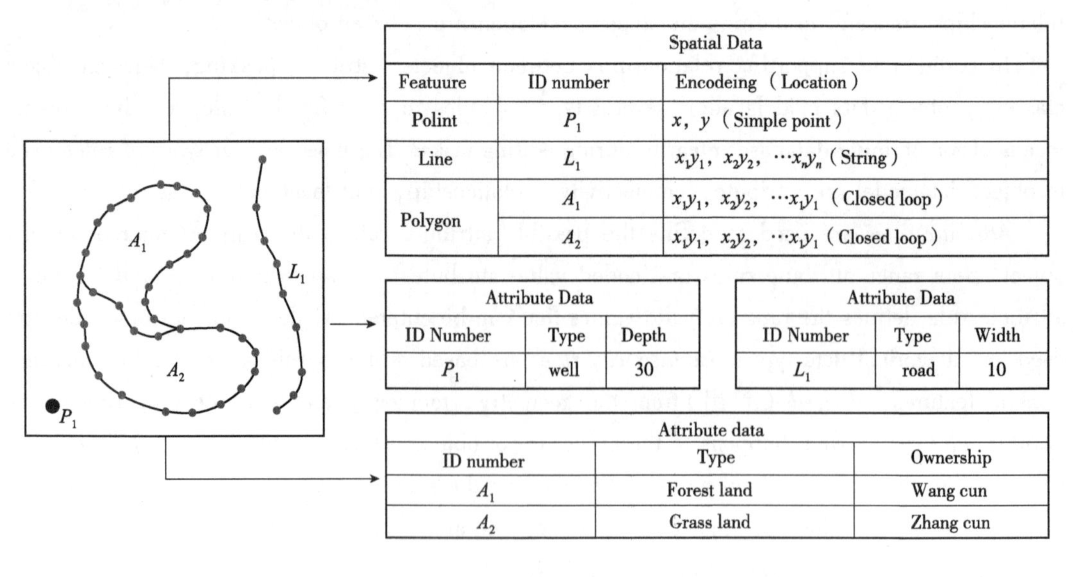

Figure 5-8 Spaghetti Data Coding of Point, Line and Polygon

①Simplicity.

②The geometry of any spatial entity is described independently of other entities.

③No topology/connectivity information is recorded.

④Points, lines and polygons are stored separately.

⑤For each polygon, a list of coordinates of points on its boundary is stored.

(8) Topologic Data Structure and Data Encoding

The most common topological data structure is the arc/node data model. This model contains two basic entities, the arc and the node. The arc is a series of points, joined by straight line segments start and end at a node. The node is an intersection point where two or more arcs meet. Nodes also occur at the end of a dangling arc. A polygon feature consists of closed chain of arcs (Figure 5-9).

Topologic data structure comprises 3 topological components which permit relationships between all spatial elements to be defined (note: does not imply inclusion of attribute data).

①Arc-Node Topology: defines relations between points, by specifying which are connected to form arcs; defines relations between arcs (lines), by specifying which arcs are connected to form routes and networks.

②Polygon-Arc Topology: defines polygons (areas) by specifying which arcs comprise their boundary.

③Left-Right Topology: defines relationships between polygons (and thus all areas) by defining starting nodes and ending nodes, which permit the left polygon and the right polygon to be specified (also the left side and the right side arc characteristics).

④Storing connectivity information explicitly allows more efficient spatial queries.

⑤Topology/Connectivity: the important criterion to establish the correctness (integrity, consistency) of geometric objects, in the applications of CAD, geographical databases, etc.

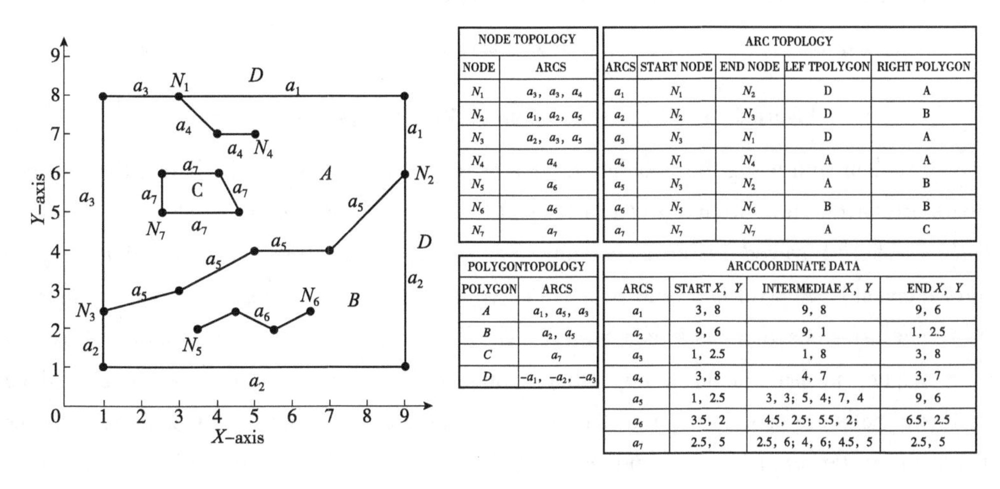

NODE TOPOLOGY	
NODE	ARCS
N_1	a_3, a_3, a_4
N_2	a_1, a_2, a_5
N_3	a_2, a_3, a_5
N_4	a_4
N_5	a_6
N_6	a_6
N_7	a_7

ARC TOPOLOGY				
ARCS	START NODE	END NODE	LEF TPOLYGON	RIGHT POLYGON
a_1	N_1	N_2	D	A
a_2	N_2	N_3	D	B
a_3	N_3	N_1	D	A
a_4	N_1	N_4	A	A
a_5	N_3	N_2	A	B
a_6	N_5	N_6	B	B
a_7	N_7	N_7	A	C

POLYGONTOPOLOGY	
POLYGON	ARCS
A	a_1, a_5, a_3
B	a_2, a_5
C	a_7
D	$-a_1$, $-a_2$, $-a_3$

ARCCOORDINATE DATA			
ARCS	START *X*, *Y*	INTERMEDIAE *X*, *Y*	END *X*, *Y*
a_1	3, 8	9, 8	9, 6
a_2	9, 6	9, 1	1, 2.5
a_3	1, 2.5	1, 8	3, 8
a_4	3, 8	4, 7	3, 7
a_5	1, 2.5	3, 3; 5, 4; 7, 4	9, 6
a_6	3.5, 2	4.5, 2.5; 5.5, 2;	6.5, 2.5
a_7	2.5, 5	2.5, 6; 4, 6; 4.5, 5	2.5, 5

Figure 5-9 Coding of Topologic Data

5. 2. 2 The Raster Data Model

Raster data models use an array of cells or pixels to represent real-world objects.

5. 2. 2. 1 Elements of Raster Data Model

Raster model is used to represent continuous numeric values and continuous *categories* (类别) are, such as digital aerial photographs, imagery from satellites, digital pictures, or even *scanned maps* (扫描地图). Nevertheless, raster data model is different from an image data model. The raster data model represents such features as a matrix/lattice of cells in continuous space. A point is one cell, a line is a continuous row of cells, and an area is represented as continuous touching cells. Raster data have an attribute table that can be joint to other tables so that they can have multiple attribute fields. Thus, it can be used to conduct spatial analysis and modeling; However, image data do not have attribute tables attached so that they have only one attribute field. It can only be used for image processing.

Raster data represents geographic data by equally spaced *discretizing* (离散化) and quantizing each raster cell. A raster cell is usually a square, but can *theoretically* (理论上) be a regular polygon that is able to fully cover an image area without leaving holes in the covered region, e. g. a triangle, *hexagon* (六角形) or rectangle (Figure 5-10).

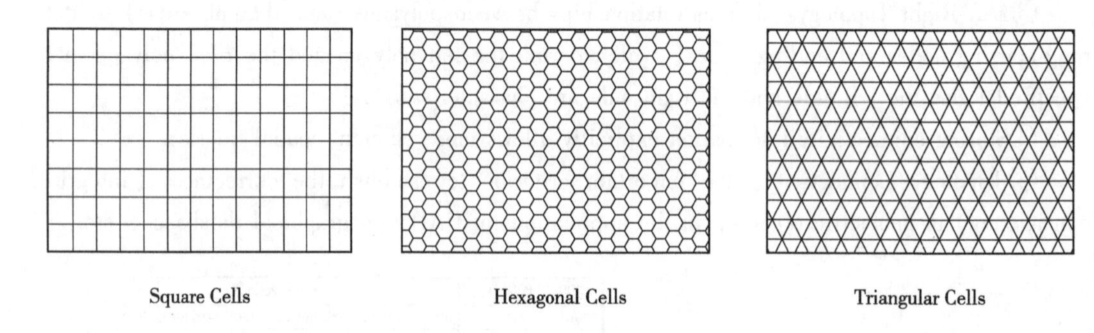

Figure 5-10 Three Most Common Regular Tessellation Types

One problem with using the popular squares or rectangles as raster cells is that the four corner neighbors are farther away from a given raster cell than the four immediate neighbors. This is more *elegantly* (优雅地) solved in hexagonal units. A raster cell is often also referred to as a pixel (picture element). A pixel can hold data value within the specified possible range or *color depth* (色深) of a raster image or raster geodata set. The data value can represent a color or gray value, depth or height, measurements or any other thematic value, such as an index to a land cover class. Raster cells are usually organized in a matrix (rows and columns). By specifying the coordinates of the raster *origin*(原点) and the *spatial resolution*(空间分辨率) of a raster cell, the spatial position of each cell within the raster grid can be easily calculated (Figure 5-11).

origin

x–location

y–location

9	4	4	4	0	5	9	9	4	4
9	5	4	0	6	0	0	7	4	6
0	7	2	7	8	9	4	7	3	8
6	3	1	1	7	8	7	3	6	1
2	2	6	7	5	7	9	0	7	4
7	6	2	8	7	8	2	8	5	8
7	8	7	3	0	9	0	0	5	2
5	8	5	5	6	5	2	2	2	1
6	2	3	4	5	6	9	0	1	4
6	9	5	1	3	6	4	4	4	1

cell size

Figure 5-11 Generic Structure for a Grid

5.2.2.2 Determination of Raster Cell Code/Value

As previously mentioned (如前所述), the raster representation divides the entire study area into a regular grid of cells. The entire study area is space-filled every location in the study area corresponds to a cell in the raster and each cell contains a single code/value. However, when different types of spatial objects (or properties) are distributed in one cell, the problem arises about how to determine the single code/value of this cell. Commonly, there are four main methods for determining cell values: *Centroid Method*, *Dominant Law*, *Importance Law* and *Percentage Breakdown Method* (中心点法、占优法、重要性法、百分比法) (Figure 5-12). Each method has been designed to represent a certain type of data better than the other. The choice of method depends on the type of data to be gridded and the analysis to be performed.

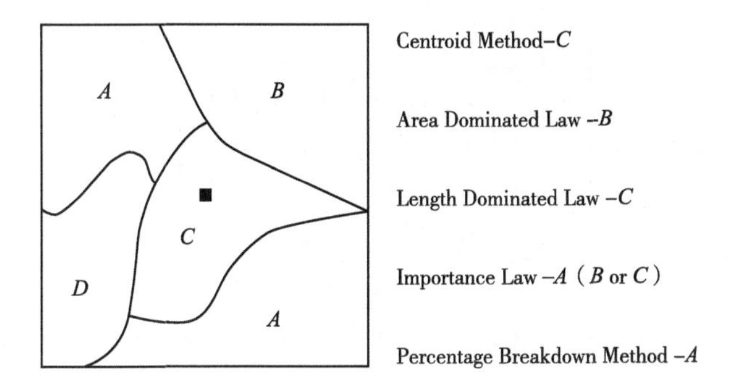

Figure 5-12 The Confirmation of the Raster Grid Code/Value

①Centroid Method: The code or value of each grid cell is determined as the corresponding attribute value of the grid center in the domain of the cell. This method can be used for any feature type, but is particularly useful in coding continuous data, such as elevation, noise movement, or *fluid flow* (液体流动). It will be at that location (the center of the cell) that a sample will be taken and a surface value will be recorded.

②Area Dominated Law: The value of each grid cell is determined as the attribute value which occupies the largest area, or also can be determined as the attribute value which occupy the largest length by the Length Dominated Law. This method is good for discrete or non-continuous data such as land cover, vegetation, or soil, where the boundaries of the objects can be defined, and their associated values assign to the cell when it occupies the majority of the cell.

③Importance Law: According to the degree of importance of different objects within the grid, the selected spatial entity of particular importance determines the value of the corresponding grid unit, even if it doesn't fill the majority of the grid cell or fall at the center of the cell.

④Percentage Breakdown Method: In this method, a cell is assigned several values (one value per feature), according to the percentage each feature occupies within the cell. This is a difficult and costly method to implement at data entry. However, it can be especially useful for statistical data (ESRI, 1998).

5.2.2.3 Raster Storage and Compression Techniques

For data with unique values for every cell, there is no way to compress the information. This is usually true for floating-point grids or continuous surfaces. The row and the column identify the location, and the value defines the attribute. The first value represents the row number and is followed by a comma. The second value is the column number. The number after the *colon* (冒号) is the value assigned to the cell. The actual storage of the cell-by-cell code in the computer may vary widely according to the implementation of the theory by the software. For example, cell-by-cell storage of the simple grid in Figure 5-13 may appear as follows:

	1	2	3	4	5	6	7	8	9	10
1	0	0	0	0	0	0	0	0	0	0
2	0	0	0	0	1	1	0	0	0	0
3	0	1	1	1	1	1	1	0	0	0
4	0	1	1	1	1	1	1	1	1	0
5	0	0	0	1	1	1	1	1	1	0
6	0	0	0	1	1	1	1	1	0	0
7	0	0	1	1	0	1	0	0	0	0
8	0	0	0	0	0	1	0	0	0	0
9	0	0	0	0	0	0	0	0	0	0

Figure 5-13 Simple Region on a Raster Grid

1, 1: 0; 1, 2: 0;...1, 10: 0

2, 1: 0;...; 2, 4: 0; 2, 5: 1; 2, 6: 1; 2, 7: 0;...2, 10: 0

3, 1: 0; 3, 2: 1;…; 3, 7: 1; 3, 8: 0;…3, 10: 0

…

9, 1: 0;…9, 10: 0

When dealing with large data sets, there are several algorithms used to compress the data. Some of these algorithms are completely *reversible* (可逆的); that is, we may recover exactly the original data sets. Others minimize the volume of the stored data by losing a (preferably) small and controlled amount of the original information. We will briefly mention four perfect-recovery compression mechanisms: *Chain coding* (the Freeman coding, 链码), *Run-length coding* (游程长度编码), *Block coding* (块状编码), and *Quadtree coding* (四叉树编码).

(1) Chain Coding

Chain codes, in some instances, we can consider a map as a set of spatially referenced objects placed on top of a background. The use of chain codes takes this point of view. The coordinate of a starting point on the border of an object (for example, a reservoir) is recorded, and then we store the sequence of *cardinal* (主要的) directions of the cells that make up the boundary. With this compression method, only the coordinates of the starting point of each line is stored.

All other subsequent points (pixels) are described by a number from 0 to 7 representing 8 possible directional positions in respect to the previous pixel. Three bits of storage would therefore be needed for each pixel of the chain code (Figure 5-14).

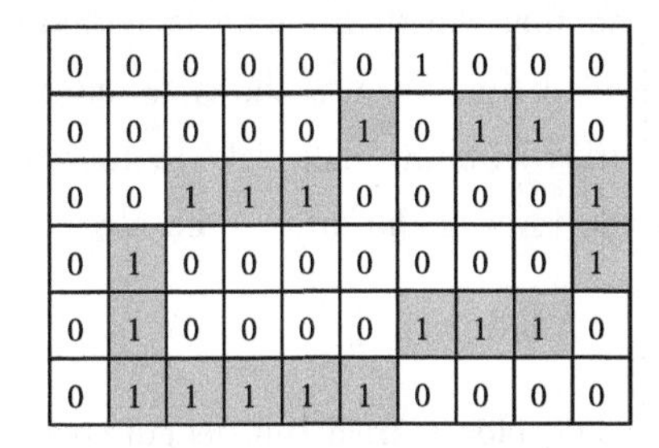

0	0	0	0	0	0	1	0	0	0
0	0	0	0	0	1	0	1	1	0
0	0	1	1	1	0	0	0	0	1
0	1	0	0	0	0	0	0	0	1
0	1	0	0	0	0	1	1	1	0
0	1	1	1	1	1	0	0	0	0

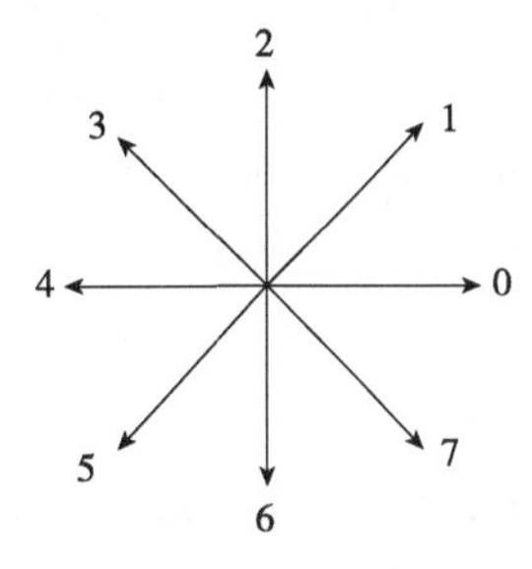

	Chain Coding
Start x, y	Encoding
1, 7	7, 0, 7, 6, 5, 4, 4, 5, 4, 4, 4, 4, 2, 2, 1, 0, 0, 1, 1

Figure 5-14 Chain Coding

This method has the characteristics as follows: ①codes can be stored using integer types; ②only boundary information is stored; ③shape analysis is possible (perimeter, directional analysis, shape turns); ④used in raster-to-vector conversion; ⑤all boundaries between regions are stored twice.

Raster															Run–length Codes
1	2	2	2	3	3	3	2	2	1	3	3	3	1	1	1 : 1, 3 : 2, 3 : 3, 2 : 2, 1 : 1, 3 : 3, 2 : 1
2	2	2	2	3	3	3	3	2	2	1	1	1	1	1	4 : 2, 4 : 3, 2 : 2, 5 : 1
2	2	2	1	1	1	2	1	2	2	3	1	2	2	2	3 : 2, 3 : 1, 1 : 2, 1 : 1, 2 : 2, 1 : 3, 1 : 1, 3 : 2
1	1	3	3	2	2	2	1	3	3	2	2	2	3	3	2 : 1, 2 : 3, 3 : 2, 1 : 1, 2 : 3, 3 : 2, 2 : 3
1	1	1	3	3	3	2	3	3	3	3	3	3	3	3	3 : 1, 3 : 3, 1 : 2, 9 : 3

Figure 5-15　Run-length Coding

(2) Run-length Coding

Run-length encoding tries to exploit the fact that many data sets have large homogeneous regions. The general encoding form is: (length, attribute value). Consider the raster data in Figure 5-15. The data attribute values at the beginning of the fifth row from the top are: 1 1 1 3 3 3 2 3 3 3 3 3 3 3 3.

In a run-length encodedversion, the original data is replaced by data pairs or *tuples* (组值). The first number in the pair is a counter, indicating how many repetitions of the second number, and the data value indicates that the number occurs starting at that point in the row. Thus, three cells in a row with data value 1 are compressed from three elements (1 1 1) to two (3 1). The data from the beginning of row 5 in our example would then become: (3 1) (3 3) (1 2) (9 3). Thus, we have three 1s, followed by three 3s, a single 2, and then nine 3s. In this case, the original data occupies 16 elements and the compressed data occupies 8 elements, for a compression factor of 50%. Note that we have assumed that the data elements including both attribute values and repeat counts in the run-length encoded file, occupy the same amount of space. The effectiveness of this compression mechanism varies with the data set. In the worst case, where there are no repeating sequences at all along the rows of the array, the algorithm will make the data set twice as large.

(3) Block Coding

Block coding is the new 2-dimension coding developed from Run-length code with square unit. It uses a technique called *medial axis transformation* (MAT) to create a data structure that stores just 3 numbers for each homogeneous block of an attribute layer. The 3 numbers are the origin (x, y) and the *radius* (半径) of each square. The origin can be the center cell or the bottom left cell. The general encoding form is: (row, column, length of the square, code) (Figure 5-16).

(4) Quad Tree Coding

The name quad tree comes from the four-fold reduction in the number of pixels in each layer. The structure is easily pictured (as well as represented in the digital database) as a tree, showing the four blocks produced at each level of *subdivision* (细分) of the squares. A polygon can be recognized as a collection of cells represented by cells (Figure 5-17).

The detailed geometric form is picked out by the combination of units of different sizes. Provided that the empty space coincides with a square at a higher level, it does not need to be repre-

0	2	2	5	5	5	5	5
2	2	2	2	2	5	5	5
2	2	2	2	3	3	5	5
0	0	2	3	3	3	5	5
0	0	3	3	3	3	5	3
0	0	0	3	3	3	3	3
0	0	0	0	3	3	3	3

1, 1, 1, 0	1, 2, 2, 2	1, 4, 1, 5	1, 5, 1, 5
1, 6, 2, 5	1, 8, 1, 5	2, 1, 1, 2	2, 4, 1, 2
2, 5, 1, 2	2, 8, 1, 5	3, 3, 1, 2	3, 4, 1, 2
3, 5, 2, 3	3, 7, 2, 5	4, 1, 2, 0	4, 3, 1, 2
4, 4, 1, 3	5, 4, 1, 3	5, 4, 2, 3	5, 6, 1, 3
5, 7, 1, 5	5, 8, 1, 3	6, 1, 3, 0	6, 6, 3, 6
7, 4, 1, 0	7, 5, 1, 3	8, 4, 1, 0	8, 5, 1, 0

Figure 5-16 Block Coding

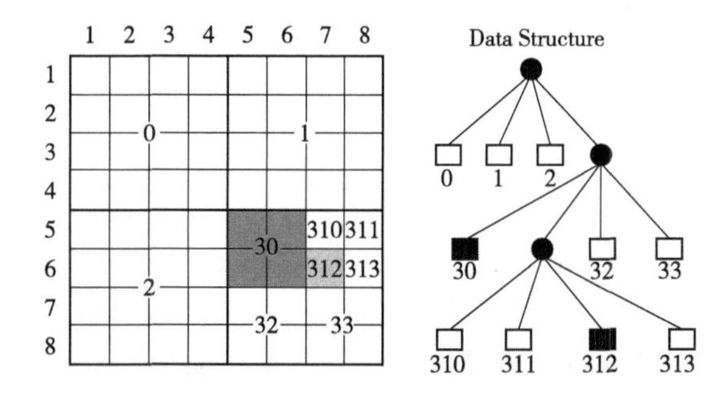

Figure 5-17 Quad Tree Coding

sented by multiple small squares. The attribute data condition is shown at the different tree levels by the (conventional) black and white coding.

For the moment, we encode locations by the numbers 0, 1, 2 and 3 to refer to the sequence of NW, NE, SW and SE blocks. Thus, 312 stands for the SE corner at level 1, the NE at level 2, and the SW at level 3.

The most efficient methods of compact representation of space are based on successive, hierarchical division of $2^n \times 2^n$ array. This is the most commonly used form of hierarchically subdividing spatial data. Subdivision continues until either a predetermined limit of resolution is reached, or all detail in the phenomenon is accounted for. The hierarchical organization allows spatial resolution to vary with the complexity of the phenomena being recorded. The lowest limit of division is the single cell.

Quad tree coding have many advantages compared to the other methods of raster representation as follows: ①easy to compute the *metric* (度量) property of polygon; ②the resolution is variable, which can represent the detail shape with a smaller spatial units; ③the transfer from quad tree to simple raster structure is easier than other compressed methods; ④able to represent the event of a complex polygon with holes.

Quad tree coding has drawbacks also. One of the drawbacks of using quad-tree is that the tree representation is data frame-dependent and not translation-invariant. The quad tree data structure is that it is not *invariant* (不变的) under translation, *rotation* (旋转), or *scaling* (缩放). So, two regions of identical shape and size may have quite different quad-tree, depending on how the square extent is defined and where the primary subdivision are drawn. This is the same as the problem found in arbitrary subdivision of datasets. Arbitrary boundaries are placed on the underlying continuous data to develop the quad tree, and the locations of the boundaries can strongly affect the resulting derived data. The quad-tree has to be rebuilt if a region changes, or is reprojected. Pattern recognition and shape analysis are more difficult, which is a problem for objects that move or change as time.

5.2.2.4 The Type of Raster Data

(1) Satellite Imagery

Remote sensingsatellite data are recorded in raster format. Spatial resolution varies: ①30 m—for Landsat 4 and 5 (using the Thematic Mapper scanner), and Landsat 7 (using Enhanced Thematic Mapper-Plus, ETM+ scanner); ②20 m—for SPOT images (Multi-spectral sensor), and 10 m—for SPOT Panchromatic sensor); ③4 m and 1 m—for IKONOS Multi-spectral and Panchromatic images respectively.

The pixel value in asatellite image represents light energy *reflected* (反射) or *emitted* (发射) from the earth's surface.

①The measurement of light energy isbased on *electromagnetic spectrum* (电磁波谱).

②*Panchromatic images* (全色影像) are comprised of a single spectral band.

③*Multi-spectral images* (多光谱图像) have multiple bands. e. g. Landsat TM has 7 bands.

④Land use, land cover and *hydrography* (水文学) can be classified from image processing system.

⑤Satellite images can be displayed in black and white or in color.

(2) Digital Elevation Models (DEM)

A Digital Elevation Model (DEM) is an ordered array of numbers that represents the spatial distribution of elevations above some *arbitrary datum* (任意起算值) in the landscape. In principle, a DEM describes the elevation of any point in a given area in the digital format and contains information of the so-called 'skeleton' lines. Skeleton lines are lines of slope reversals (drainage, crests) and breaks of slope. DEM consists of an array of uniformly spaced elevation data. DEMs are produced from: a stereo plotter (立体绘图仪) and aerial photograph with overlapping areas; satellite imagery such as SPOT stereomodel using special software.

(3) Digital Orthophotos (数字正摄影像)

They are produced from aerial photograph or other remotely sensed data. Displacement causedby *camera tilt* (相机倾斜) and terrain relief has been removed. They are geo-referenced and can *be registered with* (注册) topographic and other maps.

(4) Binary Scanned Files (二进制扫描文件)

Scanned imagescontaining values of 1 and 0. Digitized maps aretypically scanned at 300 or 400 dpi (dots per inch).

(5) Graphic Files

Maps, photographs andimages can be stored as digital graphic files. e. g. TIFF (tagged image file format), GIF (graphic interchangeable format), JPEG (joint photographic exports group), etc. GeoTIFF is a geo-referenced version of TIFF.

5.3 Comparison of Vector and Raster

5.3.1 Differences between Vector Data and Raster Data

Vector data sets andraster data sets are both important in a GIS. Each of them has its *strengths* (优势); Therefore, it's *counterproductive* (事与愿违的) to use only one of these datasets. A noticeable difference between these two data sets is the visualization of data sets. The graphical representation of raster or vector data results in *raster graphics*(栅格图) or *vector graphics* (矢量图) (Figure 5-18). In a raster data set, the attributes are associated with a raster cell. Therefore, the appearance is only changeable by modifying the color value of the raster cell. In comparison, vector data stores the information associated with the object, the appearance may be modified easily by adding graphical components to every object. In a GIS the difference between a raster and a vector data set could look like this.

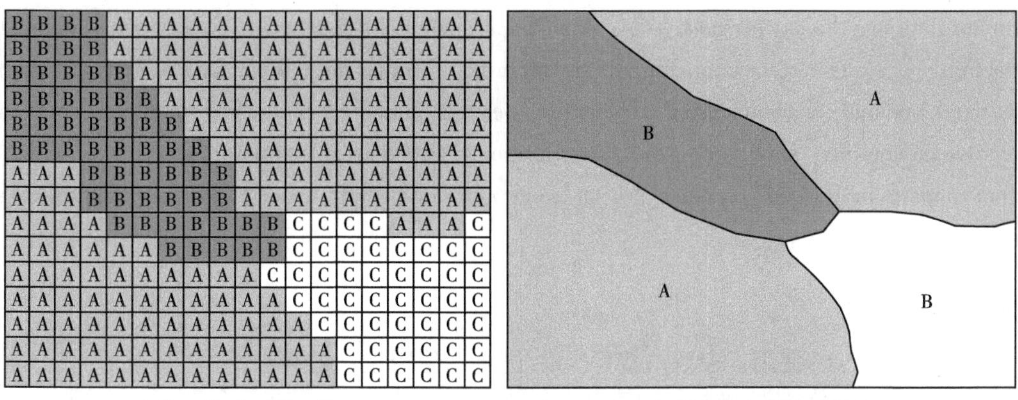

A. Visualised as Raster Data B. Visualised as Vector Data

Figure 5-18 Visualization of Data

5.3.2 Advantages and Disadvantages

Advantages of vector data are: ①Data can be represented at its original resolution and form without generalization. ② Graphic output is usually more *aesthetically* (美学上的) pleasing. ③Since most data, e. g. hard copy maps, is in vector form; no data conversion is required. ④Accurate geographic location of data is maintained. ⑤For efficient encoding of topology, more efficient operations on topological information, e. g. proximity, network analysis are required. On the other hand, the disadvantages of vector data are: ①The location of each vertex needs to be stored explicitly. ②For effective analysis, vector data must be converted into a topological structure. This is often *intensive* processing(内在的)and usually requires *extensive* (广泛的)data cleaning. As well, topology is static, and any updating or editing of the vector data requires rebuilding of the topology. ③Algorithms for manipulative and analysis functions are complex and may be intensive processing. Often this inherently limits the functionality for large data sets. e. g. a large number of features. ④Continuous data, such as elevation data, is not effectively represented in vector form. Usually substantial data generalization or interpolation is required for these data types. ⑤Spatial analysis and *filtering* (过滤)within polygons is impossible.

Correspondingly, advantages of raster data are: ①The geographic location of each cell is implied by its position in the cell matrix. Accordingly, except for an origin point, e. g. bottom left corner, no geographic coordinates are stored. ②Due to the nature of the data storage technique, data analysis is usually made in an easy and quick manner. ③The inherent nature of raster maps, e. g. one map of attribute, is ideally suited for mathematical modeling and quantitative analysis. ④Discrete data, e. g. forestry stands, is accommodated equally well as continuous data, e. g. elevation data, and facilitates the integrating of the two data types. ⑤Grid-cell system is very compatible with raster-based output devices, e. g. *plotters* (绘图仪), *graphic terminals* (图形终端).

The disadvantages of raster data are: ①The cell size determines the resolution at which the data is represented. ②It is especially difficult to adequately represent linear features based on the cell resolution. Accordingly, network linkages are difficult to establish. ③Processing associated attribute data may be *cumbersome* (难处理)if large amounts of data exist. Raster maps inherently reflect only one attribute or characteristic for an area. ④Since most input data is in vector form, data must undergo vector-to-raster conversion. Besides increased processing requirements, data integrityconcerns may arise due to the generalization and choice of inappropriate cell size. ⑤Most output maps from grid-cell systems do *not conform to* (不符合) the high-quality cartographic needs.

Vocabulary

access[ˈæksɛs] *vt.* 使用；存取；接近 *n.* 进入；使用权；通路

adjacency[əˈdʒesnsi] *n.* 毗邻；四周；邻接物

all-encompassing 包罗万象的

ambiguous[æmˈbɪɡjuəs] *adj.* 模糊不清的；引起歧义的

anchor[ˈæŋkə] *n.* 锚；抛锚停泊；靠山；新闻节目主播 *vt.* 抛锚；使固定；主持节目 *vi.* 抛锚 *adj.* 末棒的；最后一棒的

approximation[əˈprɑksəˈmeʃən] *n.* [数] 近似法；接近；[数] 近似值

arbitrarily[ˌɑrbəˈtrɛrəli] *adv.* 武断地；反复无常地；专横地

aspect[ˈæspɛkt] *n.* 方面；方向；形势；外貌

cardinal[ˈkɑrdɪnl] *n.* 红衣主教；枢机主教；鲜红色；[鸟类](北美)主红雀 *adj.* 主要的，基本的；深红色的

cognitively[ˈkɑɪɡnətɪvli] *adv.* 认知地

collinear[kəˈlɪnɪə] *adj.* [数] 共线的；同线的；在同一直线上的
constrain[kənˈstren] *vt.* 驱使；强迫；束缚
construct[kənˈstrʌkt] *vt.* 建造，构造；创立 *n.* 构想，概念
counterproductive[ˌkaʊntəprəˈdʌktɪv] *adj.* 反生产的；事与愿违的
distinct[dɪˈstɪŋkt] *adj.* 明显的；独特的；清楚的；有区别的
enabling[ɪnˈeblɪŋ] *adj.* 授权的 *v.* 使能够；授权给（enable 的现在分词）
encapsulation[ɪnˌkæpsəˈleʃən] *n.* 封装；包装
encompass[ɪnˈkʌmpəs] *vt.* 包含；包围，环绕；完成
explicitly[ɪkˈsplɪsɪtli] *adv.* 明确地；明白地
gradient[ˈgredɪənt] *n.* [数][物] 梯度；坡度；倾斜度 *adj.* 倾斜的；步行的
homogeneity[ˌhɑmədʒəˈniəti] *n.* 同质；同种；同质性(等于 homogeneousness)
inheritance[ɪnˈhɛrɪtəns] *n.* 继承；遗传；遗产
intention[ɪnˈtɛnʃən] *n.* 意图；目的；意向；愈合
interclass[ˌɪntəˈklæs] *adj.* 年级之间的；阶级之间的；组内的
interpolation[ɪnˈtəpəˈleʃən] *n.* 插入；篡改；添写
intrinsic[ɪnˈtrɪnsɪk] *adj.* 本质的，固有的
matrices[ˈmetrɪsiz] *n.* [数] 矩阵；模型；[生][地] 基质；母岩（matrix 的复数）
medium[ˈmidɪəm] *adj.* 中间的，中等的；半生熟的 *n.* 方法；媒体；媒介；中间物
node[nəʊd] *n.* 节点；瘤；[数] 叉点 *n.*（Node)人名，(法)诺德
orthophoto[ˌɔːəuˈfəutəu] *n.* 正色摄影
plotter[ˈplɒtə(r)] *n.* 绘图仪；阴谋者，策划者
polymorphism[ˌpɑlɪˈmɔrfɪzm] *n.* 多态性；多形性；同质多晶
semantics[sɪˈmæntɪks] *n.* [语] 语义学；语义论
spaghetti[spəˈgɛti] *n.* 意大利式细面条 *n.*（Spaghetti)人名，(意)斯帕盖蒂
strength[strɛŋθ] *n.* 力量；力气；兵力；优势
tangible[ˈtændʒəbl] *adj.* 有形的；切实的；实在的 *n.* 有形资产
term[təm] *n.* 术语；学期；期限；条款 *vt.* 把…叫作 *n.*（Term)人名，(泰)丁
tessellation [ˌtɛsɪˈleʃən] *n.* 棋盘形布置；棋盘花纹镶嵌；镶嵌式铺装；镶嵌细工；棋盘花纹
theoretically[ˌθiəˈrɛtɪkli] *adv.* 理论地；理论上
topology[təˈpɑlədʒi] *n.* 拓扑学；地志学；局部解剖学
undertake[ˌʌndəˈtek] *vt.* 承担，保证；从事；同意；试图
uniquely[juːˈniː kli] *adv.* 独特地；珍奇地；唯一地
vitally [ˈvaɪtəli] *adv.* 极其地；紧要地；生死攸关地

Questions for Further Study

1. What is a data model?

2. Dimensionality and property distinguish the geometric objects of point, line, and area, why?

3. Why is data model very important?
4. What are geographic data models? The importance of data models in GIS.
5. What is Topology?
6. What is TIN? What are its basic data elements?
7. What are object-oriented data and characteristic?
8. The advantages and disadvantages of the two models in GIS.
9. Distinguish between raster data model and vector data model.

Chapter 6

Creation and Maintenance of Geographic Databases

Data collection, as the most important of all the GIS tasks, is the most time-consuming and expensive. Nowadays, many methods are available to import diverse source of geographic data into GIS via data collection and data transfer. When it comes to data collection, it is important to distinguish between primary (direct measurement) and secondary (derivation from other sources) data capture for both raster and vector data types. In addition, data transfer mostly involves transferring data from other sources into digital ones and importing them into GIS [数据采集是所有 GIS 任务中最重要的一项，同时也是最费时和开销最大的。如今，有许多方法可通过数据采集和数据转换将多样化的地理数据输入到 GIS 中。数据采集时，区分栅格和矢量数据类型的原始（野外实测）与间接（从其他数据源转换）数据获取方式是很重要的。此外，数据转换主要是将其他数据源的数据转换成数字形式输入到 GIS 中].

In the fundamental process of an operational GIS, large amounts of geographic databases should be built up. *Indisputably*（毋庸置疑）, database is the most important part of GIS besides the human factor, because the collection and maintenance of databases is the most time-consuming and easily *overspent*（花销过多）in economic cost. At the same time, database is the basis of all queries, analysis, and decision making of a GIS. Essentially, all large GIS implementations store data in a *database management system* (DBMS), which is a special piece of software designed to handle multi-user access to an integrated set of data.

6.1 Data Acquisition

It is obvious that GIS will not be functional without data. Generally, more data imported, the more *versatility*（多功能；多样性）of a potential functionality GIS will have. On the implementation of GIS, the considerations of data are more important than issues of hardware or software. And considerations of data acquisition takes priority in establishing a GIS system（数据采集是建立 GIS 系统首先应考虑的问题）. Especially, collection of good-quality data plays a vital role in supplying objective information for solving problems so that some analytical understanding of the problems and hence solutions can be obtained. Meanwhile, costs of data acquisition are now considerably more than those of purchasing hardware or software themselves. Frank et al. (1991) estimated that the ratio of the costs for hardware, software and data is 1 : 10 : 100 respectively over the lifetime of GIS.

6.1.1 Classification of Data Collection

Existing data collection methods can fall into two categories: primary and secondary (Table 6-1).

Table 6-1 Summary of Primary and Secondary Data Collection

Item	Primary Data	Secondary Data
Data Sources	The real world	Libraries, remote sensing, government offices, mapping agencies, research institutes, digitizing agencies, ...
Collection Methods	Counting & measuring, questionnaires, surveying, photography, remote sensing, GPS	Networking, digitizing, scanning, database entry, photogrammetry conversion or transfer
Equipment	Measuring/surveying equipment, cameras, data loggers, positioning systems	Scanner, digitizer, computers & peripherals, image analysis equipment

Primary data is information collected directly via ground surveying, interviews, photography, RS, or GPS. According to specific needs, it provides GIS users with the most accurate and up-to-date data. Secondary data comes from earlier studies or other systems, e.g. paper maps, photos or external data. In this case, the data is input into the system via network transfer, digitizers, scanners, *stereo plotters* (立体绘图仪) or other computer *peripherals* (外部设备).

6.1.2 Methods of Primary Data Collection

6.1.2.1 Global Navigation Satellite System

The Global Navigation Satellite System (GNSS) positioning uses a set of observations such as pseudo-range of satellites, ephemeris, satellite launch time, and must also know the user *clock difference*(钟差). GNSS is a space-based radio navigation and positioning system capable of providing users with all-weather, *three-dimensional coordinate* (三维坐标) and velocity and time information anywhere on the earth's surface or in near-earth space.

The performance of GNSS is assessed with four criteria:

Accuracy(准确度): the difference between a receiver's measured and real position, speed or time.

Integrity(完整性): a system's capacity to provide a threshold of confidence and, in the event of an anomaly in the positioning data, an alarm.

Continuity(连续性): a system's ability to function without interruption.

Availability (可用性): the percentage of time a signal fulfills the accuracy, integrity and continuity.

This performance can be improved by regional satellite-based augmentation systems (SBAS), such as the European Geostationary Navigation Overlay Service (EGNOS). EGNOS improves the accuracy and reliability of GPS information by correcting signal measurement errors and providing information about the integrity of its signals.

More info. can be checked at the course of *GNSS Principles and Applications*.

6.1.2.2 Surveying

Surveying (ground surveying) is the technique of accurately determining the terrestrial or three-dimensional position of points, distances and angles. These points are often used to establish landmaps and boundaries for ownership or governmental purposes. Traditional surveying equipment includes a *tape measure* (卷尺) and *theodolite* (经纬仪) that can measure both angles and distances. Nowadays, *electronic total stations* (全站仪) have made the technological shift from optical-mechanical to fully electronic. The total station is an electronic theodolite (transit) integrated with an *electronic distance meter* (电子测距仪, EDM) to read slope distances from the instrument to a particular point.

6.1.2.3 Photography

Photography (摄影) is the art, science and practice of creating *durable* (耐久的) images by recording light or other electromagnetic radiation, either chemically by means of a light-sensitive material such as photographic film, or electronically by means of an image sensor. Typically, a lens is used to focus the light reflected or emitted from objects into a real image on the light-sensitive surface inside a camera during a timed *exposure* (曝光). The result in an electronic image sensor is an electrical charge at each pixel, which is electronically processed and stored in a digital image file for subsequent display or processing. The result in a *photographic emulsion* (感光乳剂) is an invisible *latent* (隐藏的) image, which is later chemically developed into a visible image, either negative or positive depending on the purpose of the photographic material and the method of processing. A negative image on film is traditionally used to photographically create a positive image on a paper base, known as a print, either by using an enlarger or by contact printing.

UAV (Unmanned Aerial Vehicle) aerial survey is a powerful supplement to traditional aerial photography measurement means, with the characteristics of mobility and flexibility, high efficiency and speed and accuracy, low operating costs, wide ranges of application, short production cycle and so on. UAV aerial survey offers significant advantages in the rapid acquisition of high-resolution imagery in small areas and difficult-to-fly areas. With the development of UAV and digital cameras, digital aerial photography based on the UAV platform has shown its unique advantages, and the combination of UAV and aerial photography measurement makes "UAV digital low-altitude remote sensing" a new development direction in the field of aviation remote sensing. UAV aerial photography can be widely used in the construction of major national projects, disaster emergency response and handling, land resource investigation and monitoring, digital city construction, emergency disaster relief and other fields.

6.1.2.4 Remote Sensing

Remote sensing (RS) is the measurement of physical, chemical, and biological properties of objects without direct contact (遥感是不直接接触物体而对其物理、化学和生物属性进行测量的方法). It is the technology of using aerial sensor technologies to detect and classify objects on Earth (both on the surface, and in the atmosphere and oceans) by means of *propagated* (传播) signals. The Earth's surface and atmosphere naturally radiate energy in the form of *microwaves* (微波). Microwave data can be collected by satellite microwave sensors. When *electromagnetic radiation* (电磁辐射) falls upon a surface, some of its energy is absorbed, some transmitted through the surface, and the rest reflected. Surfaces also naturally emit radiation, mostly in the form of heat. It is reflected and emitted radiation recorded either on the photographic film or digital sensor. Since the intensity and wavelength of radiation are a function of the surface in question, each surface is described as processing a characteristic spectral signature. If an instrument can identify and distinguish between different spectral signatures, then it will be possible to map the extent of surfaces using remote sensing. Therefore, remote sensing is a technology to identify and understand the object or the environmental condition through the uniqueness of the reflection or emission (Figure 6-1).

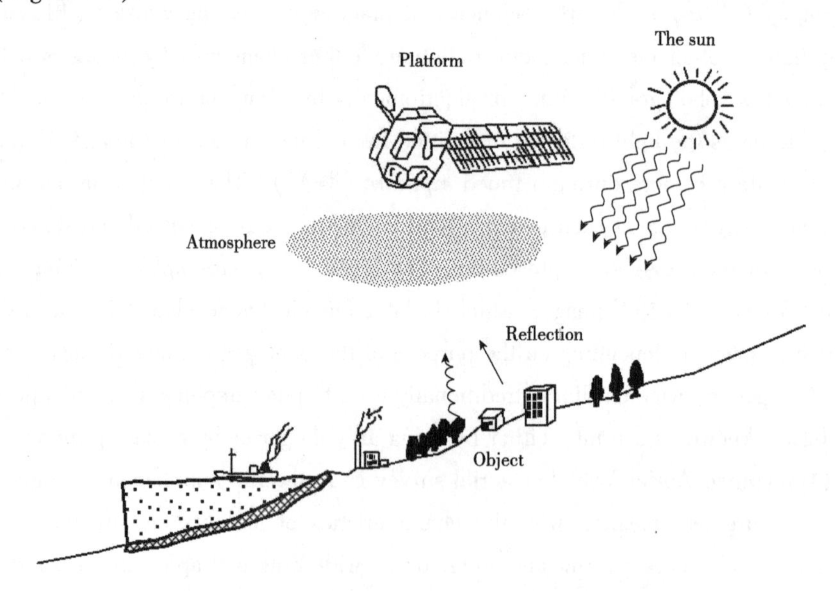

Figure 6-1 Data Collection by Remote Sensing

The *vehicle* (车辆) or carrier for remote sensors is called the platform. According to the height of RS platform, the RS can be classfied into: aerospace remote sensing, aerial remote sensing and ground sensing (按遥感平台的高度，遥感可分为航天遥感、航空遥感和地面遥感). According to the electromagnetic spectrum used, the RS can be divided into: visible reflective infrared remote sensing; thermal infrared remote sensing, and microwave remote sensing (按

所利用的电磁波的光谱段，遥感可分为可见反射红外遥感，热红外遥感和微波遥感). According to the applications in many different fields, the RS can be divided into two categories: Resources remote sensing and environmental remote sensing (按不同领域中的应用，遥感可分为资源遥感与环境遥感两大类). According to the working mode of remote sensors, the RS can be divided into passive RS and active RS (按照遥感器的工作方式，遥感可分为被动遥感和主动遥感两大类).

A remote-sensing image is defined as an image produced by a recording device free from physical or *intimate* (亲密的) contact with the object under study. This may be a map or other image obtained through various remote-sensing devices (or sensors) such as cameras, computers, lasers, *radio frequency receivers* (无线电频率接收器), *radar systems* (雷达系统), *sonar* (声波定位仪), *seismographs* (地动仪), *gravimeters* (重力计), *magnetometers* (磁传感器), and *scintillation counters* (闪烁计数器). When the image has cartographic or *bibliographic* (书目的) information added, it is defined as a remote-sensing map.

Resolutions of RS image vary in three key aspects: spatial, temporal and spectral.

Spatial Resolution (空间分辨率): It is a measure of the smallest angular or linear separation between two objects that can be resolved by the sensor. The greater the sensor's resolution, the greater the data volume and smaller the area coverage. In fact, the area coverage and resolution are inter-dependent and these factors determine the scale of the imagery. Satellite remote sensing systems typically provide data with pixel sizes in the range 0.5 m–1 km.

Temporal Resolution (时间分辨率): It, also called repeat cycle, specifies the *revisiting frequency*(重访频率) of a satellite sensor for a specific location. It refers to how often a given sensor obtains imagery of a particular area. Ideally, the sensor obtains data repetitively to capture unique discriminating characteristics of the interested phenomena.

Spectral Resolution (光谱分辨率): It refers to the dimension and number of specific wavelength intervals in the electromagnetic spectrum to which a sensor is sensitive. Narrow bandwidth of the electromagnetic spectrum in certain regions allows the discrimination of various features more easily. However, compared to the number of bands, the position of bands in the electromagnetic spectrum is important too. Since different objects emit and reflect different types and amounts of radiation, selecting electromagnetic spectrum for each application area has become critical. Remote sensing systems may capture data in one part of the spectrum (referred to as a single band) or simultaneously from several parts (multi-band or multi-spectral). The radiation values are usually normalized and resembled to give a range of integers from 0–255 for each band (part of the electromagnetic measured) for each pixel and in each image. Until recently, remote sensing satellites typically measured a small number of bands in the visible part of the spectrum. More recently a number of *hyper spectral systems* (超光谱系统) have come into operation to measure very large numbers of bands across a much wider part of the spectrum.

The principal advantages of remote sensing are the speed at which data can be acquired from large areas of the earth's surface, and the application to inaccessible areas. The major advantages of this technique over ground-based methods are summarized as follows.

①*Synoptic View*（全局性）: Remote sensing process facilitates the study of various earths' surface features in their spatial relation to each other and helps to *delineate*（描绘）the required features and phenomena.

②*Repeatability*（重复性）: The remote sensing satellites provide repetitive coverage of the earth and this temporal information is very useful for studying *landscape dynamics*（景观动态）, *phenological*（生物气候学）variations of vegetation and change detection analysis.

③*Accessibility*（可达性）: Remote sensing process makes it possible to gather information about the area where it is not possible to do ground survey like in mountainous areas and foreign areas.

④Time Saving: Since information about a large area can be gathered quickly, the techniques save time and human efforts. It also saves the time of fieldwork.

⑤Cost Saving: Remote sensing especially when conducted from space is an intrinsically expensive activity. Nevertheless, cost-benefit analysis demonstrates its financial effectiveness, and much speculation or developmental remote sensing activity can be justified in this way. It is a cost-effective technique as again and again fieldwork is not required and also a large number of users can share and use the same data.

6.1.2.5 LIDAR

LIDAR is a method for measuring distances (ranging) by illuminating the target with laser light and measuring the reflection with a sensor. Differences in laser return times and wavelengths can then be used to make digital 3D representations of the target. It has terrestrial, airborne, and mobile applications.

The term LIDAR was originally aportmanteau of light and radar. It is now also used as an acronym of "light detection and ranging" and "laser imaging, detection, and ranging". LIDAR sometimes is called 3D laser scanning, a special combination of a 3D scanning and laser scanning.

LIDAR is commonly used to make high-resolution maps and is mostly used insurveying applications in geodesy, geomatics, archaeology, geography, geology, geomorphology, seismology, forestry, atmospheric physics, laser guidance, airborne laser swath mapping (ALSM), and laser altimetry. The technology is also used in control and navigation for someautonomous cars.

6.1.3 Methods of Secondary Data Collection

Geographic data collection from secondary sources is the process of creating raster and vector files and databases from maps, photographs, and other hard-copy documents.

(1) Scanning

A scanner is a device that converts hard-copy analog media to digital images by scanning successive lines across a map or document and recording the amount of light reflected from a local data source (Figure 6-2). The differences in reflected light are normally *scaled into* (扩展到) bilevel black and white (1 bit per pixel), or multiple gray levels (8, 16, 32 or 64 bits). Color scanners output data into 8-bit red, green, and blue color hands. The spatial resolution of scanners varies widely from as little as 200 dpi (8 dots per millimeter) to 2400 dpi (96 dots per millimeter) and beyond. Most GIS scanning is in the range 400 - 900 dpi (16 - 40 dots per millimeter).

Figure 6-2 Scanner

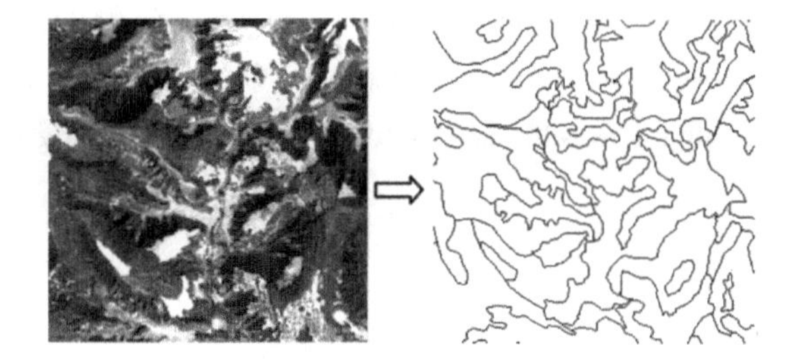

Figure 6-3 Raster to Vector Conversion

(2) Conversion Between Raster and Vector Data

Vectorization (矢量化) is the process of converting raster data into vector graphics (vector data). It is popular in applications of *computer-aided design* (CAD) drawings (blueprints etc.), GIS, graphic design and photography. In GIS, satellite or aerial images are vectorized to create maps. Many different algorithms exist for the vectorization process and each gives different results, as vector representations are more abstract than pixels (Figure 6-3).

Rasterization is the task of taking an image described in a vector graphics format (shape) and then converting it into a raster image (pixels or dots) for output on a video display or printer, or for storage in a bitmap format. The term "rasterisation" in general can be applied to any process by which vector information can be converted into a raster format. Rasterization indicates the popular rendering algorithm for displaying 3D shapes on a computer. It is currently the most popular technique for producing real-time 3D computer graphics. Real-time applications need to respond immediately to user input, and generally need to produce frame rates of at least 24 frames per second to achieve smooth animation.

Compared with other rendering techniques such as ray tracing, rasterization is extremely fast. However, rasterization is simply the process of computing the mapping from scene geometry to pixels without prescribing a particular way to compute the color of those pixels. Shading, including programmable shading, may be based on physical light transport, or artistic intent. The process of rasterizing 3D models onto a 2D plane for display on a computer screen is often carried out by fixed function hardware within the graphics pipeline. This is because there is no motivation for modifying the techniques for rasterisation used at render time and a special-purpose system allows for high efficiency.

(3) Digitizing

Digitizing or digitization is the representation of anobject, image, sound, document or signal [usually an *analog signal* (模拟符号)] by a discrete set of its points or samples. The result is called digital representation or, more specifically, a digital image, for the object, and digital form, for the signal. Digitizing indicates capturing an analog signal in digital form.

Manual digitizing is still the simplest, easiest, and cheapest method of capturing vector data from existing maps. In manual-digitizing techniques, a map or aerial photograph is placed on a digitizing table (Figure 6-4) and a pointing device [called a cursor, *puck* (橡胶圆球), or mouse] is used to record coordinates of features to be extracted from the map. The digitizing table electronically encodes the position of the *cursor* (光标). Tracing the map features with the cursor can be time-consuming and error-prone. Recent advances in scanning hardware and software have made scanning a feasible alternative to manual digitizing for some applications.

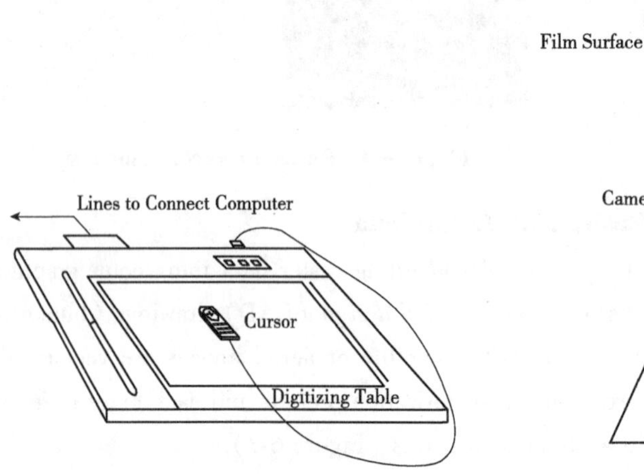

Figure 6-4 A Digitizing Table

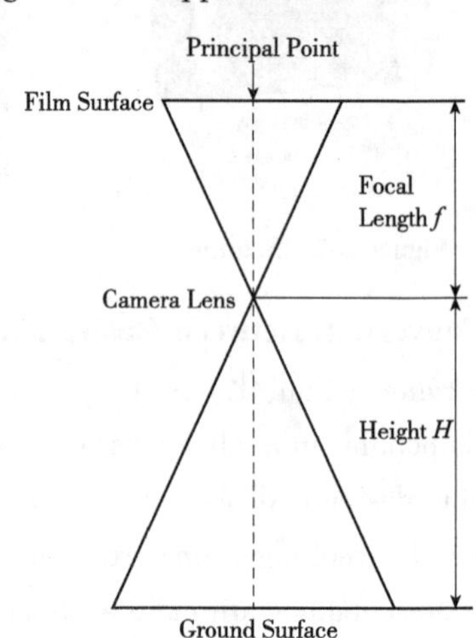

Figure 6-5 Frame of Photogrammetry

(4) Photogrammetry

Photogrammetry is the practice of determining the geometric properties of objects from photographic images; it is used to capture measurements from photographs and other image sources in GIS. The workflow of digital photogrammetry is data entry, processing and product generation. Data can be obtained directly from sensors or by scanning secondary sources (Figure 6-5). Unfortunately, the complexity and high cost of equipment have restricted its use to large-scale primary data capture projects and specialized data capture organizations.

(5) Data Import or Transfer

A decision has to be made onwhether to build or buy parts or all of a database at the start of a GIS project. This section focuses on how to import or transfer data into a GIS that has been captured by others. Some datasets are free, but many of them are sold as commodities from a variety of outlets including internet websites.

The best access to geographical data is to search geographic data geo *portals* (门户网站), such as US NSDI Clearinghouse (http://www.fgdc.gov), Geography Network (http://www.geographynetwork.com), etc (Table 6-2).

Table 6-2 Selected Websites Providing Free Geographic Datasets

Source	URL	Description
Global Mapping	http: //www. iscgm. org/cgi-bin/fswiki/wiki. cgi	A set of consistent GIS layers covering the whole globe at 1km resolution including: transportation, elevation, drainage, vegetation, administrative boundaries, land cover, land use and population centers. Produced by the *International Steering Committee* (国际指导委员会) on Global Mapping
SRTM	http: //srtm. csi. cgiar. org/	Approx. 90 m (3 arc-second) resolution elevation data from the Shuttle Radar Topography Mission for the whole world
Global Multi-Resolution Topography	http: //www. marine-geo. org/portals/gmrt/	Gridded elevation at approximately 100 m resolution, covering terrestrial and sea-floor topography
Natural Disaster Hotspots	http: //sedac. ciesin. columbia. edu/data/collection/ndh	A wide range of geographic data on natural disasters [including volcanoes, earthquakes, *landslide* (滑坡), flood and 'multi-hazards'] with hazard frequency, economic loss etc.
USGS Land Cover Institute	http: //landcover. usgs. gov/landcoverdata. php	Great set of links to almost all land cover datasets. Links here include most of the datasets below, and many more *esoteric* (机密的) data such as river observations, aquifers data and ocean color information
Atlas of the Biosphere	http: //www. sage. wisc. edu/atlas/maps. php	Raster maps of environmental variables including soil pH, potential evapotranspiration, average snow depth and many more
Mineral Resources Data System	http: //tin. er. usgs. gov/mrds/	Vector data of *mineral resources* (矿产资源) across the world including names, locations, descriptions, geological characteristics etc.
UNEP GEOdata	http: //geodata. grid. unep. ch/	A wide range of data from the United Nations Environment Programme including nighttime lights, *pollutant emissions* (污染物排放), commercial shipping activity, protected areas and administrative boundaries
Natural Earth	http: //www. naturalearthdata. com/	Includes countries, *disputed areas* (有争议的地区), first-order admin (departments, states etc.), populated places, urban polygons, parks and protected areas and water boundaries. Available at multiple levels of detail
Global Land Use Dataset	http: //www. sage. wisc. edu/iamdata/	Gridded data at 0.5 degree resolution showing population density, potential natural vegetation, *cropland extent* (农田程度), *grazing land extent* (牧场程度), built-up land extent, crop extent (for 18 major crops) and land suitability for cultivation
Gridded Population of the World	http: //sedac. ciesin. columbia. edu/gpw/	Includes raw population, population density, both historic, current and predicted
Open Flights	http: //openflights. org/data. html	Airport, airline and route data across the globe. Data is provided as CSV files which can be easily processed to produce GIS outputs. Data includes all known airports, and a large number of routes between airports

（续）

Source	URL	Description
China Dimensions Data Collection	http：//sedac. ciesin. columbia. edu/data/collection/cddc	GIS data including administrative regions，census data linked to maps and agricultural data
Diamond Bay Data	http：//www. dbr. nu/data/geo/	Chinese counties，*census statistics*（人口普查统计数据）and Digital Chart of the World China GIS layers

Data can be transferred between systems via direct read into memory or an intermediate file format. Given the high cost of creating databases，many tools have been developed to move data between systems and to reuse data through open application programming interfaces（APIs）. Many GIS software systems are now able to read directly AutoCAD DWG and DXF，Micro station DGN，and Shape file，VPF，and many image formats. Unfortunately，direct read support can only easily be provided for relatively simple product-oriented formats. Geographic data can be encoded in many different ways（formats），but no single format is appropriate for all tasks and applications. The most efficient way to convert data between systems is through a common intermediate file format. And this intermediate file format normally refers to a spatial data transfer standard. Designed for exchange purposes，complex formats as SDTS require more advanced processing before they can be viewed.

6.2 Data Inputting and Editing

GIS data contains non-spatial data and spatial data. Real world data must be acquisition and converted to digital format in order to be analyzed and manipulated by a GIS. This process generally includes data inputting，data editing，error correction，geocoding，and data conversion，access and integration.

6.2.1 Data Inputting

Geometric data and attribute data are input by the following methods：

①Direct data acquisition by surveying or remote sensing. Vector data can be measured with digital survey equipment such as electronic total stations or analytical photogrammetric plotters. Raster data are sometimes obtained from remote sensing data.

②Digitization of existing maps. A map can be simply defined as a graphic representation of the real world. This representation is always an abstraction of reality. Because of the infinite nature of our Universe it is impossible to capture all of the complexity found in the real world. For example，topographic maps abstract the 3D real world at a reduced scale on a two-dimensional plane of paper. Maps are used to display both cultural and physical features of the environment on the earth. Standard topographic maps show a variety of information including roads，land-use classification，elevation，rivers and other water bodies，political boundaries，and the identification of houses and other types of buildings. The earth is curved（弯曲的）but the map is flat. The earth is big but over a small area it looks flat. So this might be true for a map of a small area. In details it cannot be true, but the issue is how far from the accuracy the map would be. We are not talking of map accuracy here, just the problem of representing a curved earth on flat paper. The only way to get around this problem is to use *a spherical map* (球面映射) or specifically a globe. The process of placing a part of the earth on a flat surface is called Projection. All flat maps have *distortion* (失真)(Figure 6-6).

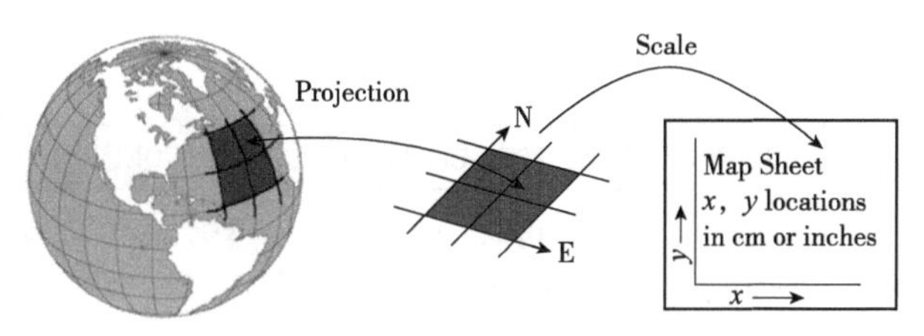

Figure 6-6 Mapping Round Earth on Flat Sheet Leads to Distortion

Existing maps can be digitized with a scanner or tablet digitizer. Raster data are obtained from a scanner while vector data are measured by a digitizer. In GIS, raster data and vector data are frequently converted to vector data and raster data respectively which are called raster/vector conversion and raster/vector conversion respectively.

A process of input of geographic data is as Figure 6-7 shown.

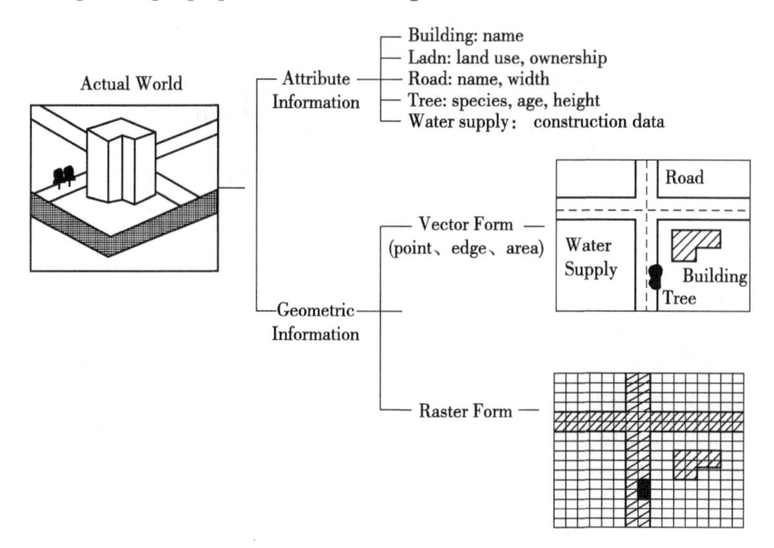

Figure 6-7 Representation of Attribute Information and Geometric Information

6.2.2 Data Editing

Editing is the process of adding, deleting and changing data.

6.2.2.1 Format Transformation

Data format transformation mainly includes two aspects:

①Raster to raster. GIS raster data are stored in various formats from a standard file-based structure to *binary large object* (BLOB) data stored directly in a relational database management system (RDBMS). The conversion between raster data formats can be implemented in different GIS systems. Popular GIS data formats are as Table 6-3 shown.

Table 6-3 Popular GIS Data Formats

Raster	Common Format	ADRG, BSQ, BIP, BIL, BMP, CADRG, CIB, DPX, DRG, ECRG, ECW, ERS, FITS, GeoTIFF, GIF, HDRi, IMG, JNG, JPEG/JFIF, JPEG2000, MrSID, PPM, PGM, PBM, PNM, PNG, RAW, SGI, TGA, TIFF, WBMP, WEBP, XBM, XCF, XPM
	Transfer Format	BIL, ECW, ERS, FST, GeoTIF, GIS, LAN, GOE, HD, TIF/TFW
Vector	Common Format	AI, CDR, CGM, DBF, DGN, DLG, DWG, DXF, EVA, EMF, Gerber, GML, GeoJSON, GeoMedia, IGS, HPGL, HVIF, IGES, ISFC, PGML, PLT, PRN, SHP, SHX, SVG, TAB, TIGER, VPF, VML, WMF, XAR
	Transfer Format	E00, GEN, MIF/MID, SHP, SIMA, WMF / EMF, XPS
Hybrid	(Raster & Vector)	CGM, PDF, PS, RasterDWG, RVD, SVG, SVGZ, SWF, XAML

②Raster to vector or vector to raster. Different raster formats and vector formats can be converted each other in different systems. Currently, nearly all GIS software gives advanced control tool over conversions, adds conversion from vector to raster formats and raster to vector formats. Vector to raster conversion is shown as Figure 6-8.

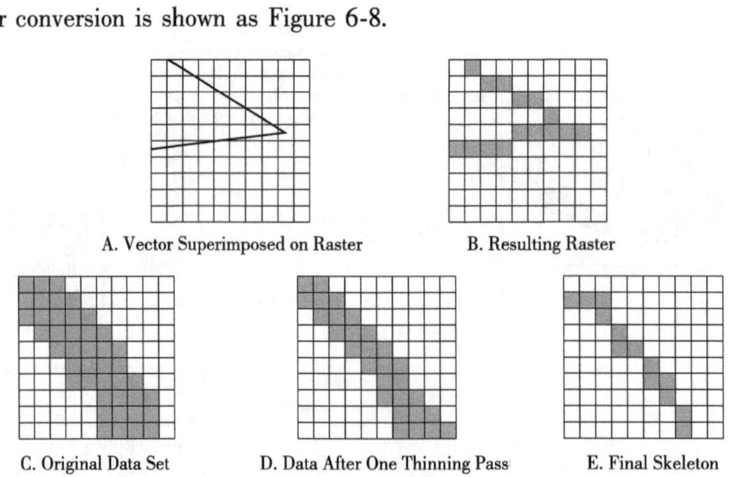

Figure 6-8 Vector to Raster Conversion

6.2.2.2 Data Transformation and Processing

Due to the variety of data sources, original spatial data are often different from data from users' own information system in the aspects of data structure, organization and presentation. Therefore, data transformation and processing are needed, such as project transformation, data restructure, data clipping, registration and extracting.

(1) Project Transformation

Map projectionis a systematic transformation of the latitudes and longitudes of locations on the surface of a sphere or an *ellipsoid* (椭球体) into locations on a plane. It normally involves a mathematical model that transforms the locations of features on the earth's surface to locations on a two-dimensional surface. Because the earth is three-dimensional, some method must be used to depict the map in two dimensions. Therefore such representations distort some parameter of the earth's surface, be it distance, area, shape, or direction. There are a variety of map projections, but all are generally of three basic types: the *azimuthal*, *conical and cylindrical projections* (方位投影、圆锥投影和圆柱投影). For example, the transverse Mercator projection is a variant of the cylindrical projection (横轴墨卡托投影是圆柱投影的一种变体).

For data with projection information, projection needs to be defined to complete the integration and exchange of data. Data with different *mapsheet* (图幅) projection need to be transformed to specific projection; the process of which is called project transformation — to transform one projection to another projection often requires transformation of projection type, projection parameter and ellipsoid (投影变换是将一种地图投影转换为另一种地图投影, 主要包括投影类型、投影参数或椭球体的改变). Most GIS soft wares use transformation of normal solution to transform different projections and offer functions that support common projection transformation.

(2) Data Restructuring

There are many sources of spatial data, such as maps, *engineering drawings* (工程图), *planning graphics* (规划图), photographs, *digital line graphics* (DLG, 数字线划图), aerial and remote sensing images, and so on. Therefore different sources of spatial data lead to different structure of spatial data. So, we need to converse data structure according to practical applications. Conversion between different data structures includes vector-to-raster conversion and raster-to-vector conversion.

(3) Data Clipping, Registering and Extracting

Data clipping (数据裁剪) is to clip an area from the whole data, in order to obtain data that we need and reduce the amount of calculation. *Data splicing* (数据拼接) is to put adjacent spatial data together and become a complete data. Research area may cover several areas of adjacent spatial data stored disparately, so it is necessary to *splice* (拼接) the spatial data together. *Data extracting* (数据提取) is to select eligible data from attribute table in order to form a new data layer. To use SQL expression is an efficient way to select data.

6.2.2.3 Image Registration and Rectification

Image registration (图像配准) is the process of transforming different sets of data into one coordinate system. Data may be multiple photographs, from different sensors, periods of time, or viewpoints. *Image rectification* (图像纠正) is a transformation process used to project two-or-more images onto a common image plane. It corrects image distortion by transforming the image in to a standard coordinate system. A kind of exact numerical approach to rectification is the transformation of map projection. While approximation approach to rectification is rubber sheeting.

Rubber Sheeting (橡皮拉伸) is the process of adjusting the coordinates of all data points in a dataset so that the known positions in the dataset coincide more correctly with each other. The process of stretching, shrinking, and changing the orientation of the interconnections between points and objects so that they retain their interconnectivity or topological relationships(Figure 6-9).

In particular, data conversion to a reference coordinate system requires changes in scale, rotation and translation (缩放、旋转和平移). The simplest and most commonly used method of a transformation is scaling, which is magnification or reduction in one or both axes. Translation linear movement along one or both axes and rotation shifts the pixel output around the origin (Figure 6-10).

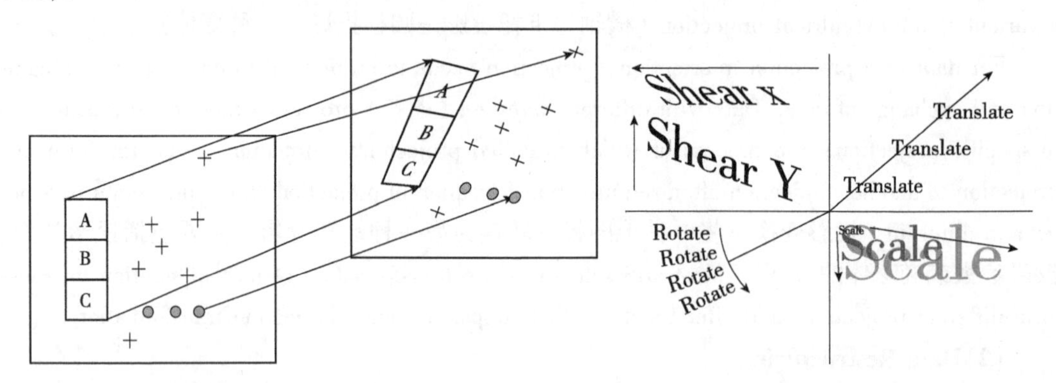

Figure 6-9 Rubber Sheeting

Figure 6-10 Scaling, Rotation, Translation in Relation to the Origin

Both registration and rectification require some forms of coordinate transformation. Transformation equation allows the reference coordinates for any data file location to be precisely estimated; to transform from one coordinate space to another.

6.2.2.4 Coordinate Transformation

Different kinds of coordinates are used to position objects in a 2D or 3D space. Spatial coordinates (also known as global coordinates) are used to locate objects either on the earth's surface in a 3D space, or on the earth's reference surface (ellipsoid or sphere) in a 2D space.

Two common types of coordinate systems are used in GIS: ①A global or spherical coordinate system such as latitude-longitude. These are often referred to as geographic coordinate systems (GCS). ②A projected coordinate system (PCS) such as *universal transverse Mercator* (UTM, 通用横轴墨卡托投影), *Albers Equal Area* (亚尔勃斯等积投影), or Robinson, provides various mechanisms to project maps of the earth's spherical surface onto a 2D *Cartesian* (笛卡儿的) coordinate plane. Projected coordinate systems are referred to as map projections.

Map and GIS users are mostly confronted in their work with transformations from one 2D coordinate system to another (Figure 6-11). This includes the transformation of polar coordinates delivered by the survey or into Cartesian map coordinates or the transformation from one 2D Cartesian (x, y) system of a specific map projection into another 2D Cartesian (x, y) system of a defined map projection.

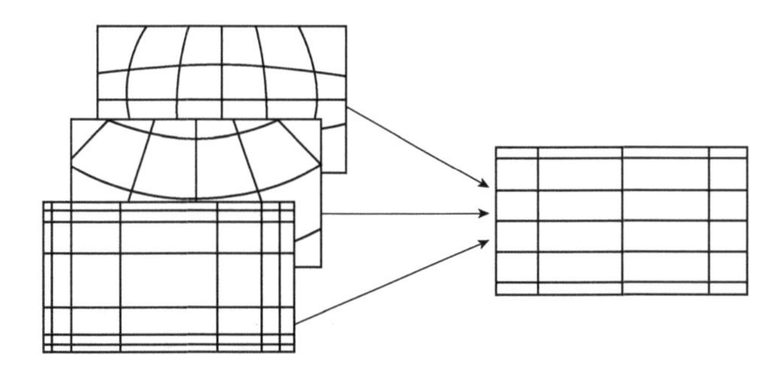

Figure 6-11 Integration of Spatial Data into One Common Coordinate System

6.2.2.5 Edge Matching

Edge matching(边缘匹配) is simply the procedure to adjust the position of features that extends across typical map sheet boundaries. Theoretically data from adjacent map sheets should meet precisely at map edges. In other words, Edge matching is an editing procedure to ensure that the joining map sheets that cross adjacent map sheets have the same edge locations (Figure 6-12).

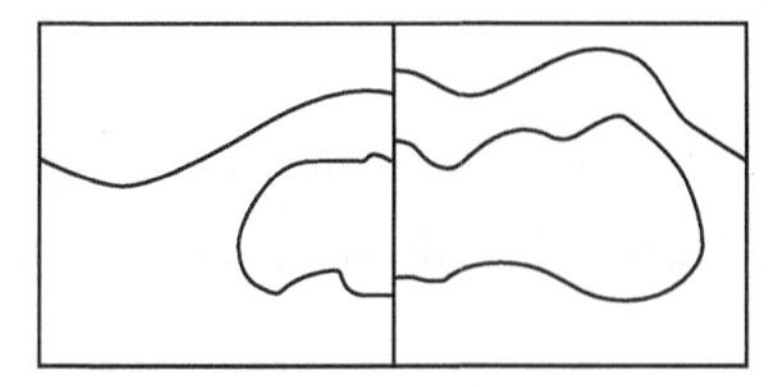

A. Original Two Map Sheets Brought Together

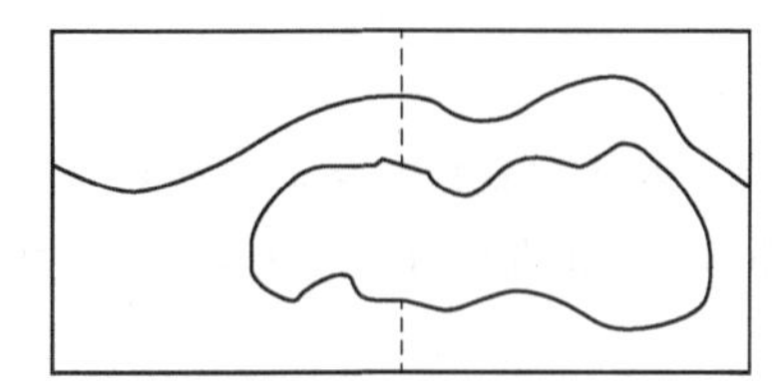

B. Derived Single Sheet with Edge Matching Showing Discrepanciesor Discontinuity

Figure 6-12 Edge Matching

Most edge matching methods, including automated or manual procedures, are based on stereo or triplet image pairs, and the results are merged together if there are more images.

6.2.2.6 Topology Creation

A GIS *topology* (拓扑) is a set of rules and behaviors that model how points, lines, and polygons share coincident geometry. Topology has long been a key GIS requirement for data management and integrity. In general, a topological data model manages spatial relationships by representing spatial objects (point, line, and area features) as an underlying graph of topological primitives-nodes, faces, and edges. These primitives, together with their relationships to one another and to the features whose boundaries they represent, are defined by representing the feature geometries in a planar graph of topological elements. Topology is fundamentally used to ensure data quality of the spatial relationships and to aid in data compilation. Creating topology is usually an *iterative process* (迭代过程) because it is seldom possible to resolve all data problems during the first pass and manual editing is required to make corrections (Figure 6-13).

Topological relationships describe relationships between objects in space. Important topological relationships are explained below:

①*Proximity* (关联性) describes how close or how far away two (or more) objects are-the

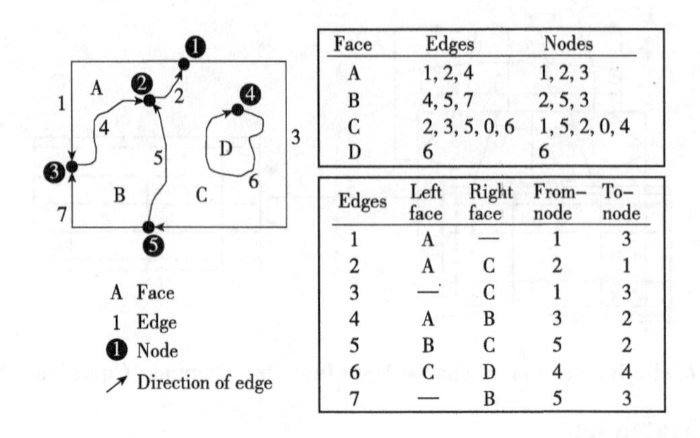

Face	Edges	Nodes
A	1, 2, 4	1, 2, 3
B	4, 5, 7	2, 5, 3
C	2, 3, 5, 0, 6	1, 5, 2, 0, 4
D	6	6

Edges	Left face	Right face	From-node	To-node
1	A	—	1	3
2	A	C	2	1
3	—	C	1	3
4	A	B	3	2
5	B	C	5	2
6	C	D	4	4
7	—	B	5	3

Figure 6-13 Topology elements and relationships

Northwest A&F University is near Xi'an city.

②*Connectivity*（连通性）means how two objects are linked with each other–the Shenyang city is connected with a subway.

③*Adjacency*（邻接性）explains whether two objects are next to each other or not–John's seat is next to Mike's.

④*Membership*（包含性）means whether an object belongs to a particular group or not–Yangling city, located in Shanxi province, is a typical High-tech agricultural industry demonstration zone of China.

⑤*Orientation*（方向性）describes the location and direction of an object–the Teaching Building is located 50 meters north of the school *canteen*（食堂）.

6.2.2.7 Data reduction

Data reduction is the transformation of numerical or alphabetical digital information derived empirical or experimentally into a corrected, ordered, and simplified form（数据化简是把从经验或实验中提取出的数值或按字母顺序排列的数字信息转化为校正后的、有序的和简化的形式）. It is the process of minimizing the amount of data（Figure 6-14）that needs to be stored in a data storage environment; data reduction can increase storage efficiency and reduce costs. When information is derived from instrument readings there may also be a transformation from *analog*（模拟）to digital form. When the data are already in digital form, the 'reduction' of the data typically involves some editing, scaling, coding, sorting, *collating*（整理）, and producing tabular summaries. When the observations are discrete but the underlying phenomenon is continuous then smoothing and interpolation are often needed. Often the data reduction is undertaken in the presence of reading or measurement errors（数据化简过程中常伴随着读数或测量误差）. Some idea of the nature of these errors is needed before the most likely value may be determined.

Data reduction is a method of reducing storage needs by *eliminating redundant data*（消除数据冗余）, and it can be achieved using different types of technologies, such as: data archiving works by filing infrequently accessed data to secondary data storage systems（数据归档：将不常访问的数据备案到另一数据存储系统中）; data compression reduces the size of a file by remo-

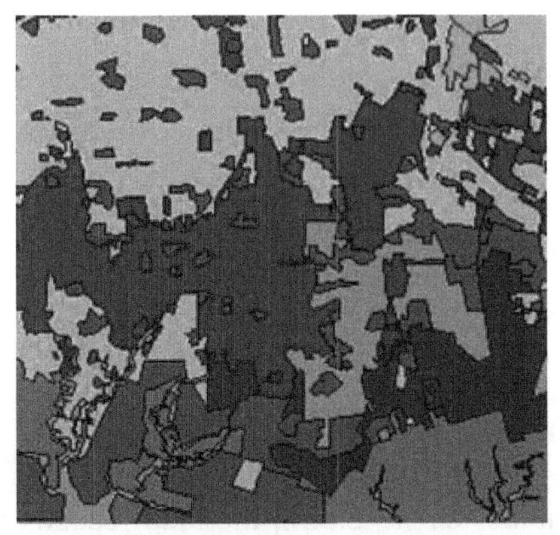

A. Minimum Mapping Unit = 1 hm^2

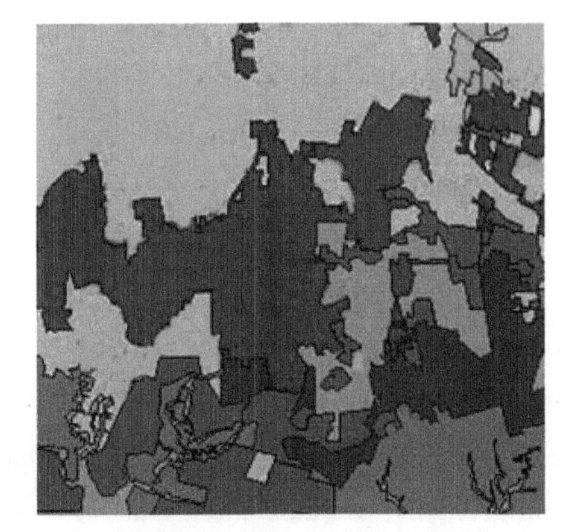

B. Minimum Mapping Unit = 9 hm^2

Figure 6-14 Data reduction-Small Area Elimination

ving redundant information from files so that less disk space is required (数据压缩：通过从文件中移除冗余信息来减小文件的大小而使得占用较少的磁盘空间).

6.2.2.8 Data Generalization

Data generalization means simplification of map information, so that information remains clear and uncluttered when map scale is reduced (数据综合是指在地图比例尺缩小时对地图信息的简化以使地图信息保持清晰和整洁) (Figure 6-15). Usually it involves a reduction in details, a resampling to larger spacing, or a reduction in the number of points in a line. It is done manually by a cartographer usually, but increasingly *semi-automated* (半自动化) and even automated methods have been used, particularly in conjunction with a GIS.

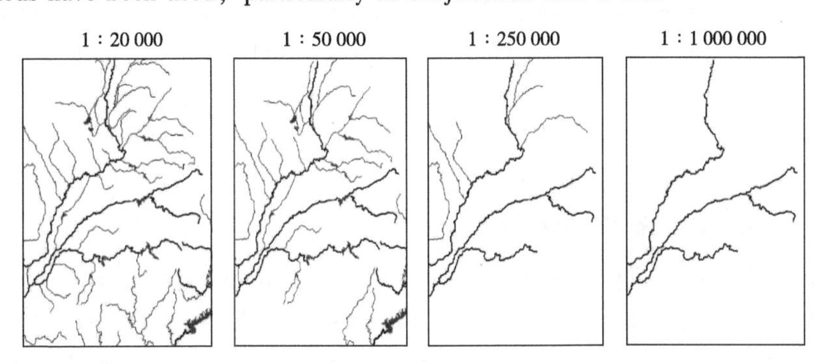

Figure 6-15 Map Generalization

Generally, there are several data generalization methods such as:

①Selection: Map generalization is designed to reduce the complexities of thereal world by strategically reducing *ancillary* (附属的) and unnecessary details. One way that these ancillary and unnecessary data can be reduced is through the selection process. For example, a cartographer often *retain and highlight*(保留并突出) certain elements that are the most necessary or appropriate while ignore the other lesser elements entirely.

②Simplification: Simplification is a technique where shapes of retained features are altered to enhance visibility and reduce complexity. Smaller scale maps have more simplified features than larger scale maps because they simply exhibit more areas. In simplification, the important characteristics are determined and unwanted details are eliminated to retain clarity on a map when its scale has been reduced.

③Combination: When considering simplification, two separate elements can be combined into one.

④Smoothing: Smoothing is yet another way of simplifying the map features, it is the process of reducing the angularity of line work to result in a smoother appearance, or removing the small variations in an image to *reveal* (揭示) the global pattern or trend.

⑤Enhancement: Enhancement can be a valuable tool in aiding the map reader. It is often employed to raster data by the cartographer to improve appearance or usability by highlighting specific details and making these specific features more *detectable* (检测). *Such operations can include contrast stretching, edge enhancement, filtering, smoothing, and sharpening* (增强操作包括对比度拉伸、边缘增强、滤波、平滑和锐化).

6.2.3 Common Errors and Editing

GIS users need to have tools to transform spatial data of various types into digital format. A *transcription error* (转录错误) is a specific type of data entry error commonly made by human operators or by *optical character recognition* (OCR, 光学字符识别) programs because of irremovable noise and incomplete original maps, which result in a large amount of manual work with resultant inefficiencies in time and cost. As well, manual data input is *labor intensive* (劳动密集型), *tedious* (乏味的) and *error-prone* (易错的), thus it may involve many kinds of errors, mistakes and misregistration. Therefore, further effort should be applied to obtain data of high-quality and reliability.

Error correction (误差校正, 纠错) is often done to provide corrections for mistakes or error through both equipment and human operators.

6.2.3.1 Problems with Digitizing Maps

Problems will arise in digitizing maps since most maps are not drafted for the purpose of digitizing. Another reason lies in that paper maps are unstable. As the map is removed from the digitizing table, the reference points need to be re-entered when the map is placed on the digitizing table again. On the other hand, if the map has stretched or shrunk *in the interim* (在此期间), the newly digitized points will be slightly off from the previously digitized points. Errors have occurred on these maps and entered into the GIS database as well. The level of error in the GIS database is directly related to the error level of the source maps.

①Cartographic Errors: Sometimes, in order to displayspecific information, the cartographer often transfer the necessary data from actual location to another mapping locations. Therefore, the displayed information on a map does not always accurately record actual location information. For example, when a railway, stream and highway all go through a narrow mountain pass, the pass

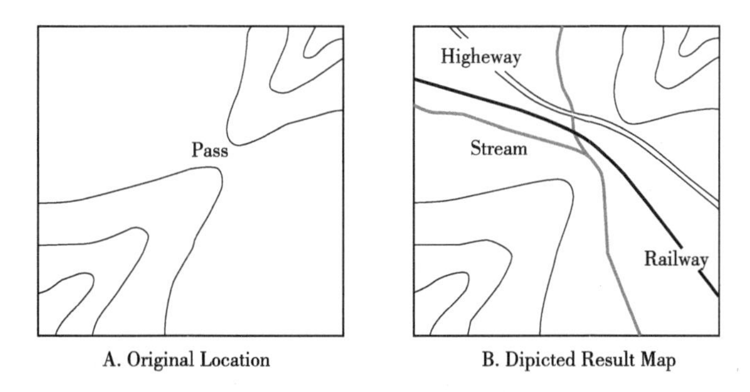

Figure 6-16 Cartographic Errors

may actually be depicted wider than its actual size to allow for these three symbols to be drafted in the pass (Figure 6-16).

②Edge Matching Errors: Discrepancies across map boundaries can cause discrepancies in the total GIS database. For example, when two maps are placed next to each other, the roads or rivers are not fully collocated.

③User Errors: *Overshoots* (过伸), *undershoots* (不及), *spikes* (锋尖), dead end, weird polygon, edge shift, attribute errors arise due to the *fatigue* (疲劳), *boredom* (厌倦) or inattention (疏忽大意) from cartographer (Figure 6-17).

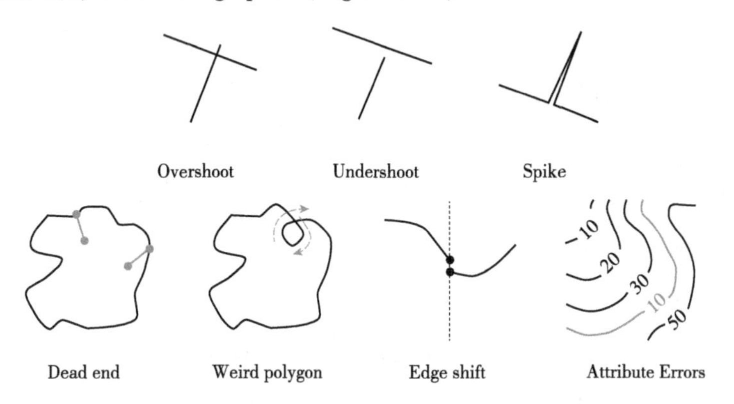

Figure 6-17 Common User Errors

6.2.3.2 Topological Errors

As Figure 6-18 shown, except for attribute errors, common topological errors with geometric features are listed as: *unclosed polygons* or *leaking polygons*(未闭合多边形), *gaps between polygons* (多边形间隙), *overlapping polygons* (多边形重叠), *dangling nodes* (悬挂节点), *pseudo nodes* (伪节点), etc (Figure 6-18).

Other topological errors arise between layers such as: line features from one layer do not connect with those of another layer (一个图层内的线要素不能与其他图层内的线要素衔接), point features of one layer do not align with line features of another layer (一个图层内的点要素不能与其他图层内线要素构成一条直线) (Figure 6-19).

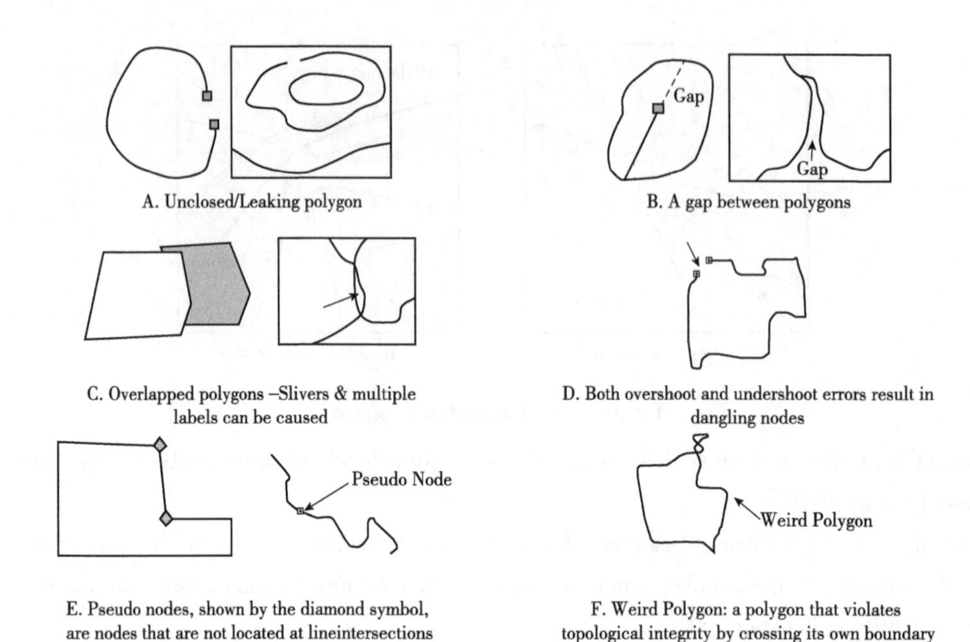

A. Unclosed/Leaking polygon

B. A gap between polygons

C. Overlapped polygons –Slivers & multiple labels can be caused

D. Both overshoot and undershoot errors result in dangling nodes

E. Pseudo nodes, shown by the diamond symbol, are nodes that are not located at lineintersections

F. Weird Polygon: a polygon that violates topological integrity by crossing its own boundary

Figure 6-18 Topological Errors with Geometric Features

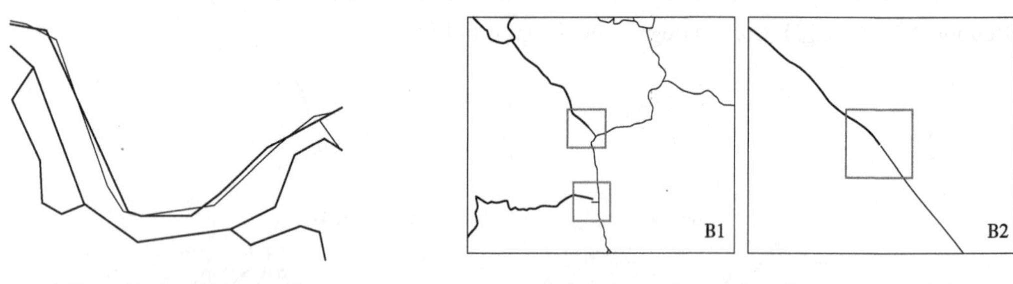

A. The outline boundaries of two layers, one shown in a thicker line and the other a thinner line, are not coincident at the top

B. Line features from one layer do not connect perfectly with those from another layer at end points. (B2) is an enlargement of the top error in (B1).

Figure 6-19 Topological Errors Between Layers

6.2.3.3 Error Correction Procedures

GIS package must be topology-based and able to detect, display, and remove errors. Some errors can be corrected automatically, e. g. small gaps at line junctions, overshoots and sudden spikes in lines. Their error rate depends on the complexity of the map, and is high for maps with small areas.

Error correction starts with building topology. Geo-database model has about 25 topology rules for point, line, and area features. Topological editing is a four-step process: ①create new topology – setting a fuzzy tolerance; ② validate topology – check automatically topological errors; ③fixing errors–editing operation; ④rebuilding topology. The process will be performed over and over again until all meets the defined topology rules.

The *fuzzy tolerance* (模糊容差) is an extremely minimum distance used to resolve inexact intersection locations due to limited arithmetic precision of computers.

Editing include: dangles-remove undershoots and overshoots; *duplicate arcs* (双重弧段)-select and delete redundant arcs; wrong arc direction-change relative positions of beginning and ending nodes; pseudo nodes-remove pseudo node by joining arcs on either side; label errors-add missing labels or remove redundant labels; *reshaping arcs* (弧段整形)-move, add, or delete points; *extend/trim* (延展/缩短) lines; delete/move features; *reshape features* (要素整形); *split* (分割) lines and polygons; create features from existing features, and so on.

6.2.4 Geocoding

The *Geocoding* (地理编码) is the important function of GIS. Geocoding, also known as address matching, refers to the process of creating spatial coordinate relationships from statistical data or address information.

Geocoding is a series of coding in order to identify the point, line, surface location and attributes in a specific associated geographic coordinate system (often expressed aslatitude and longitude). It will record the quantized data by the suitable way about the attributes of entities and set coordinate data structure in the storage device of a computer with a pre-established classification system.

6.2.5 Data Conversion, Access and Integration

Different softwares adopt different data formats, which makes data sharing inconvenient. In order to solve this problem, methods of *data format exchange* (数据格式转换), directly data access, and *data interoperation* (数据互操作) can be adopted.

As one of the general functions, data format exchange can be implemented in all GIS software, although the level of implementation is different. It is a mode of converting data to the needed format by the specialized data conversion program, and this is the main way to realize data sharing among current GIS software. In order to exchange data with other software, most GIS software makes the exchange format of hamming codes. For example, the format of E00 in ArcInfo, the format of Shape in ArcView and the format of Mif in MapInfo (为了实现与其他软件的数据交换，许多 GIS 软件均制定有明码交换格式，如 ArcInfo 的 E00 格式、ArcView 的 Shape 格式、MapInfo 的 Mif 格式).

Many GIS supports conversion algorithms that enable them to accept data in other formats.

Data integration (数据集成) in GIS involves combining data from different disparate sources, formats, *spatial and temporal scale* (时间和空间尺度), characteristics in the logical or physical organic, so that the GIS can provide a unified view of the data for data sharing. Therefore, these different datasets can reasonably be displayed on the same map and their relationships can sensibly be analyzed. Data integration is the indispensable function of GIS software widely used, and it is becoming increasingly important in merging systems of different companies or *consolidating* (巩固) applications to provide a comprehensive view of these datasets. The later initiative is often called a *data warehouse* (数据仓库).

6.2.6 New Trends in Data Collection and Data Editing

(1) Data Collection

UAV (unmanned aerial vehicle) based data acquisition has flourished, and in the future, the technology of data acquisition will be more intelligent and diverse.

(2) Data Editing

Image classification continues to develop toward the direction of automation and intelligence, and fully automatic computer classification and mapping will be vigorously developed under the push of AI.

(3) Data Transmission

The advent of 5G technology, massive amounts of data and cloud technology will bring historic changes to the transmission of data and the distribution of information.

6.3 Creating and Maintaining Geographical Databases

6.3.1 Database

A database is a *repository* (仓库) capable of storing large amounts of data. It comes with a number of useful functions: ①the database can be used by multiple users at the same time (concurrency); ②the database offers a number of techniques for storing data and allows to use the most efficient one (storage); ③the database allows to impose rules on the stored data, which will be automatically checked after each update to the data (integrity); ④the database offers an access to use data manipulation language, which allows to perform all sorts of data extraction and data updates (query facility); ⑤The database will try to execute each query in the data manipulation language in the most efficient way (query optimization).

A database provides three views of data (Figure 6-20):

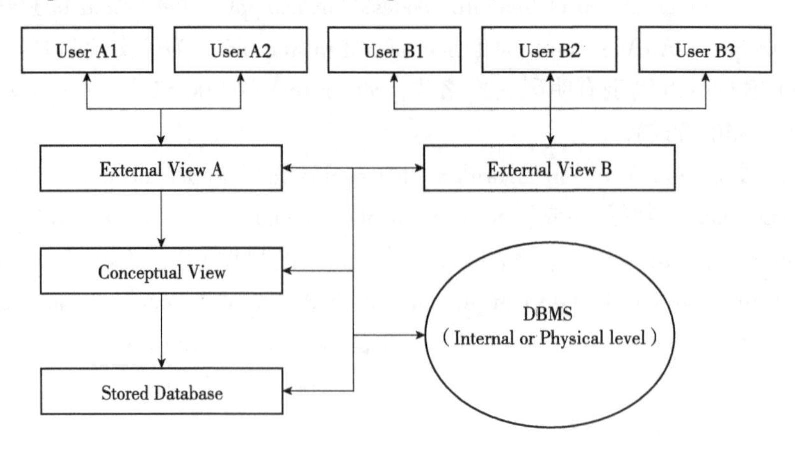

Figure 6-20 View of Database

①Internal View or Physical Level: the internal organization of data inside a DBMS, which is normally not seen by the user or applications developer.

②Conceptual View: the primary means by which the database administrator builds and manages the database. It provides the *synthesis* (综合) of all the external views.

③External View: what the end-users see. Even a single database can have any number of views to different users and applications.

6.3.2 Database Management System

6.3.2.1 Concept of DBMS

A database management system (DBMS) is designed to manage a large body of information. Data management involves both defining structures for storing information and providing mechanisms for manipulating the information. In addition, the database system must provide the safety of the stored information, despite *system crashes* (系统崩溃) or attempts at unauthorized access. If data are to be shared among several users, the system must avoid possible *anomalous* (多态的) results due to multiple users concurrently accessing the same data.

ADBMS is a set of software programs that allows users to create, edit and update data in database files, store and retrieve data from these database files. Data in a database can be added, deleted, changed, sorted or searched by using a DBMS.

6.3.2.2 Uses of DBMS

Four major uses of a DBMS are:

①Database Development: Database software packages like Microsoft Access, Lotus Approach allow end users to develop the database they need. However, large organizations with client/server or mainframe-based system usually place control of enterprise-wide database development in the hands of database administrators and other database specialists.

②Database Interrogation: The *database interrogation* (数据库的审查) capability is a major use of DBMS. End users can interrogate a DBMS by asking for information from a database using a query language or a report generator. They can receive an immediate response in the form of video displays or printed reports.

③Database Maintenance: Database maintenance is when a database is created after the work is called database maintenance. This database maintenance process is accomplished by *transaction* (事务) processing programs and other end-user application packages within the support of the DBMS. End-users and information specialists can also employ various utilities provided by a DBMS for database maintenance.

④Application Development: DBMS packages play major roles in application development. End-users, system analysts and other application developers can use the fourth-generation programming language and built-in software development tools provided by many DBMS packages to develop *custom application programs* (自定义应用程序).

6.3.2.3　Main Capabilities of DBMS

①Data Load: DBMS provides tools to load data into databases. Simple tools are available to load standard supported data types (e.g. character, number, and date) in well-structured formats.

②Non-standard Data Formats: It can be loaded by writing custom software programs that convert the data into a structure that can be read by the standard loaders.

③Indexes: An index is a data structure used to speed up searching. All databases include tools to index standard database data types.

④A Query Language: One of the major advantages of DBMS is that they support a standard data query/manipulation language called SQL (structured/standard query language).

⑤Security: A key characteristic of DBMS is that they provide controlled access to data. This includes restricting user access to all or part of a database. For example, a casual GIS user might have read-only access to just part of a database, but a specialist user might have read and write (create, update, and delete) access to the entire database.

⑥Controlled Update: Updates to databases are controlled through a transaction manager responsible for managing multi-user access and ensuring that updates affecting more than one part of the database are coordinated.

⑦Backup and Recovery: It is important that the valuable data in a database are protected from system failure and incorrect (accidental or deliberate) update. Software utilities are provided to back up all or part of a database and to recover the database in the event of a problem.

⑧Database Administration Tools: The task of setting up the structure of a database (the schema), creating and maintaining indexes, tuning to improve performance, backing up and recovering, and allocating user access rights is performed by a *database administrator* (DBA). A specialized collection of tools and a user interface are provided for this purpose.

⑨Applications: Modern DBMS are equipped with standard, general-purpose tools for creating, using and maintaining databases. These include applications for designing databases and for building user interfaces for data access and presentations (forms and reports).

⑩*Application Programming Interfaces* (APIs): Although most DBMS have good *general-purpose* (通用的) applications for standard use, most large, specialist applications will require further customization using a commercial *off-the-shelf* (现成的) programming language and a DBMS programmable API.

6.3.2.4　Types of DBMS

①Relational: A DBMS is said to be a Relational DBMS or RDBMS if the database relationships are treated in the form of a table. There are three keys on relational DBMS: relation, domain and attributes. A network means it contains fundamental constructs sets or records. Sets contain one to many relationships, records contain fields statistical table that is composed of rows and columns used to organize the database and its structure, and it is actually a two dimension array in the computer memory. Some popular DBMS includes Oracle, Sybase, Ingress, Informix, Microsoft

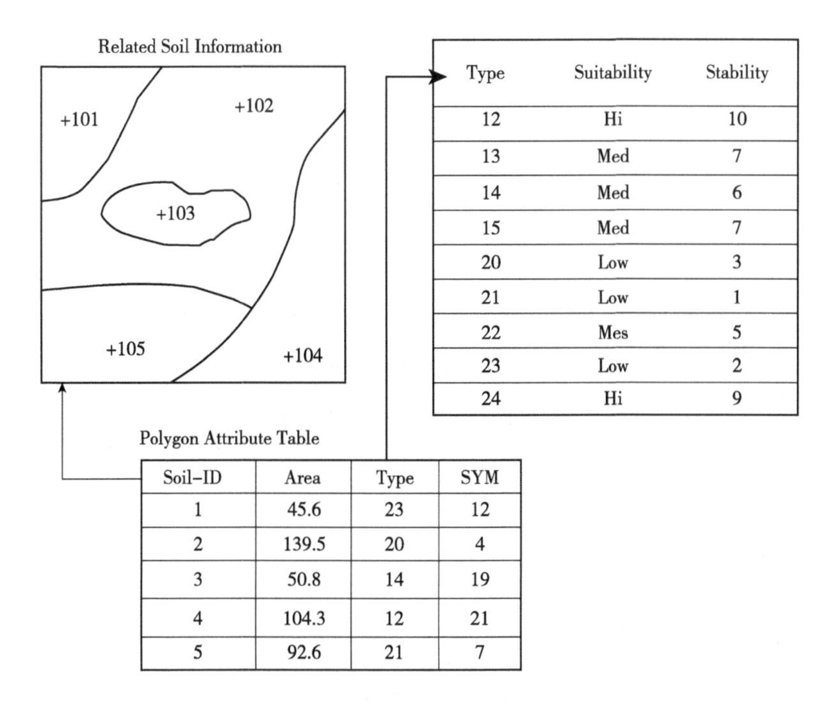

Type	Suitability	Stability
12	Hi	10
13	Med	7
14	Med	6
15	Med	7
20	Low	3
21	Low	1
22	Mes	5
23	Low	2
24	Hi	9

Soil-ID	Area	Type	SYM
1	45.6	23	12
2	139.5	20	4
3	50.8	14	19
4	104.3	12	21
5	92.6	21	7

Figure 6-21 Example of Relational Model

SQL Server, and Microsoft Access. An example of relational model is shown as Figure 6-21.

②Hierarchical: Hierarchical. A database is hierarchical if the relationships between data in the database are established in such a way that one data item is used as a *subordinate* (下级) to another data item.

Here "subordinate" means that items have parent-child relationships among them. Direct relationships exist between any two records that are stored consecutively. The data structure "tree" is followed by the DBMS to structure the database. No backward movement is possible/allowed in thehierarchical database. Hierarchical data model was developed by IBM in 1968 and introduced in information management systems. This model is like a structure of a tree with the records forming the nodes and fields forming the branches of the tree. In the hierarchical model, records are linked in the form of an organization chart. An example of hierarchical model is shown as Figure 6-22.

③Network: A DBMS is called a network DBMS if the relationships between data in the database are of the many-to-many type. Thus the structure of a network database is extremely complicated because of these many-to-many relationships in which one record can be used as a key of the entire database. The structure of such a DBMS is highly complicated, however it has two basic elements, i. e. records and sets to *designate* (指定) many-to-many relationships. Mainly high-level languages such as Pascal, COBOL and FORTRAN etc. are used to implement the records and set structures. An example of network model is shown as Figure 6-23.

④ODBMS: Object databases are different from relational databases and belong together to the broader database management system. Able to handle many new data types, including graph-

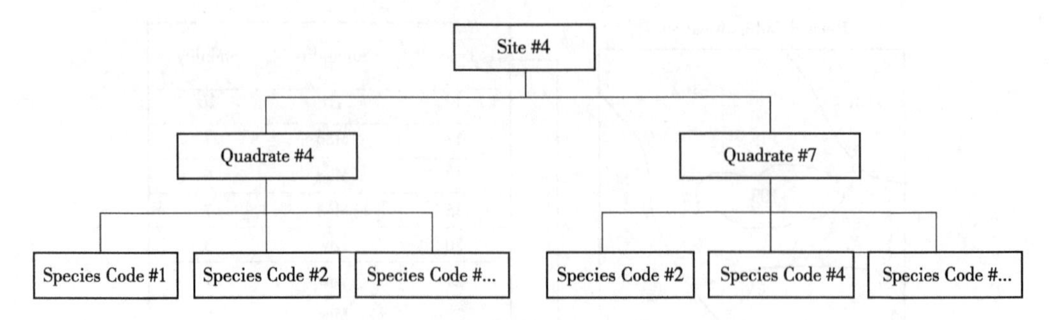

Figure 6-22 Example of Hierarchical Model

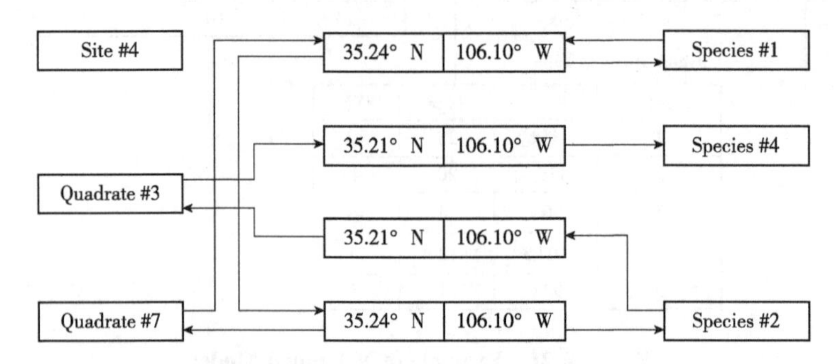

Figure 6-23 Example of Network Model

ics, photographs, *audio*（音频）, and video, object-oriented databases represent a significant advance over their other database cousins.

Object database management systems（ODBMS）were initially designed to address several weaknesses of RDBMS. These include the inability to store complete objects directly in the database. Focused primarily on business applications such as banking, human resource management, and *stock control and inventory*（库存管理和存货）, RDBMS were never designed to deal with rich data types. An example of object-oriented model is shown as Figure 6-24.

6.3.3 Database Management

GIS data consist of spatial data and thematic attribute data. They are characterized by large volume of data, unstructured, and related to spatial topological relationships（GIS 数据由空间数据和属性数据组成，它们的特点是数据量大、非结构化，并与空间拓扑关系有关）. Spatial database system manages GIS data via layer and frame（空间数据库系统对空间数据进行分幅和分层管理）.

GIS datasets are managed in the *working layer*（工作层）and *workspace*（工作区）, which is the unit for data organization or data processing for user's specific need. The workspace in GIS software specifies the default location or a folder that contains one or more shape file, geodatabase, feature dataset of the map which is composed by a series of working layer. In the Workspace, available layers are displayed in a table that includes critical information about each layer. Working layer, called layer in some software, is the data processing unit in the real sense

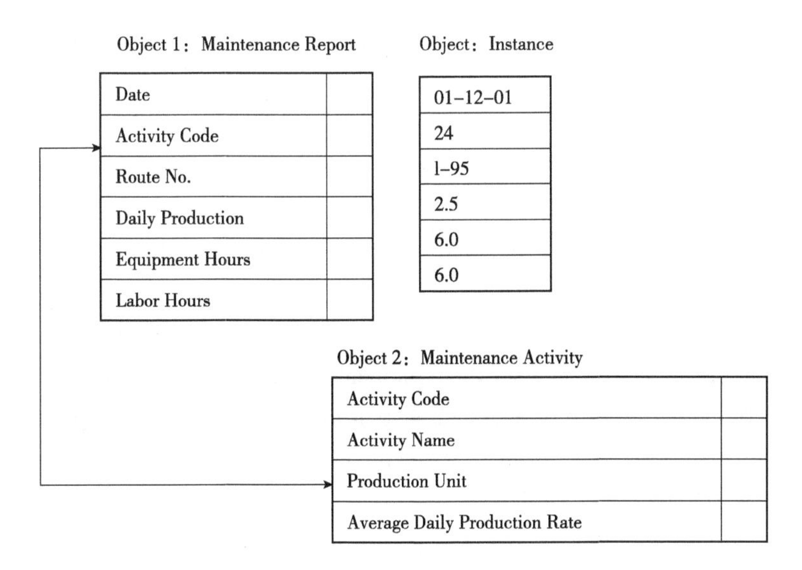

Figure 6-24 Example of Object-oriented Model

(工作层是真正意义上的数据处理单元，在一些软件中工作层被称为图层).

A data frame is a map element that defines a geographic extent, a page extent, a coordinate system, and other display properties for one or more layers in GIS software. A dataset can be represented in one or more data frames. In the data view, only one data frame is displayed at a time; while in layout view, all a map's data frames can be displayed at the same time. Since the 1990s, GIS databases based on the object-oriented spatial data model of the spatial schema have the function of organizing seamless and borderless data(无缝数据). In this way *framing management* (分幅管理) is shown as framing or partitioned indexes on seamless image, meeting the needs of users to operate and retrieval specific regional and thematic layer.

GIS data management must function as establishing *map database*(图库). In a seamless map mode, database management function can meet the needs of the specific regional and thematic layer operation and retrieve by the effective framing (partition) and spatial index hierarchy (在无缝地图模式下，数据库管理职能通过有效的分幅/分区、分层空间索引，以满足用户对具体的局部区域和专题层进行操作、检索).

6.3.4 Data Retrieval and Query

Data retrieval, in database management, involves extracting the wanted data from a database. The two primary forms of the retrieved data are reports and queries. In order to retrieve the desired data, the user presents a set of *criteria* (标准) by a query. Then the Database Management System (DBMS) or software for managing databases, selects the demanded data from the database. The retrieved data may be stored in a file, printed, or viewed on the screen.

Spatial data query, based on the ordered data set, is to find geography objects from spatial database which all meet the *attribute constraints* (属性约束) and *space constraints* (空间约束). Spatial query is to find and describe the specific features of data set, according to expression for the specific element. In essence query is to find the spatial data and attribute data set to meet some condition (查询的实质是查找满足条件的空间数据与属性数据集).

Query language allows users to interact directly with the database software in order to perform information-processing tasks using data in a database. It is usually an easy-to-use computer language that relies on basic words such as SELECT, DELETE, or MODIFY. Using query language and a computer keyboard, users put in commands that instruct the DBMS to retrieve data from a database or update data in a database.

Structured Query Language (SQL) is one query language widely used to perform operations by using relational databases. Remember that relational databases are composed of tables with rows and columns. SQL can be used to retrieve information from related tables in a database or to select and retrieve information from specific rows and columns in one or more tables. One of the keys to understanding how SQL works in a relational database is to realize that each table and column has a specific name associated with it. In order to query a table, users specify the names of the tables (indicating the rows to be displayed) and the names of the columns to be displayed.

Three key elements of SQL are "SELECT" (the column names to be displayed), "FROM" (indicates the table name from which column names will be derived) and "WHERE" (describes the condition for the query).

More information can be check at Chapter 7.

6.3.5 Data Updating

GIS data updating means to update high practicality of data or change out-of-date data in GIS database. It can achieve the goal of the maintaining reality of spatial information in the database and improve the data accuracy. At the same time, updated data can be stored in historical database for retrieval and served for the time series analysis, historical status recovery, the decision management and research services. Therefore GIS data updating is not simply to remove and replace, but to record the history (GIS 数据更新不是简单的删除和替换，更新的同时还要记录历史). In essence, GIS data updating is the process of spatial entity state changes. In other words, data updating includes two states: the real situation of geographic entities into the database, the history of geographic entities into the database.

6.3.6 Spatial Index

A spatial index is a special access method used to retrieve data from within the data-store. Spatial indexes allow users to treat data within a data-store as existing within a two dimensional context. A spatial index is a grid divided into a number of rectangles or cells. All cells have the same width and height. All cells, together, map a single large rectangle. Two fields are required from any record to map the record to the grid-effectively the (x, y) coordinate pair.

Common spatial index methods include: Grid, *Z*-order (curve), Quad tree, *Octree* (八叉树), UB-tree, R-tree, m-tree, and so on. Let's take Grid, Quad tree, R-tree for example.

①Grid: A grid in the context of a spatial index is a regular tessellation of a manifold or 2-D surface that divides it into a series of contiguous cells, which has a unique identifier and used for spatial indexing purposes. A wide variety of such grids have been proposed or are currently in use, including grids based on "square" or "rectangular" cells, triangular grids or meshes, *hexagonal grids* (六边形格网), grids based on diamond-shaped cells, and possibly more.

②Quad Tree: A quad tree is a tree data structure in which each internal node has exactly four children. Quad trees are most often used to partition a two-dimensional space by *recursively subdividing* (递归细分) it into four quadrants or regions. The regions may be "square" or "rectangular", or may have "arbitrary" shapes.

③R-tree: Typically the preferred method for indexing spatial data. Objects (shapes, lines and points) are grouped by the *minimum bounding rectangle* (MBR). Objects are added to an MBR within the index that will lead to the smallest increase in its size. An example of R-tree index is shown as Figure 6-25.

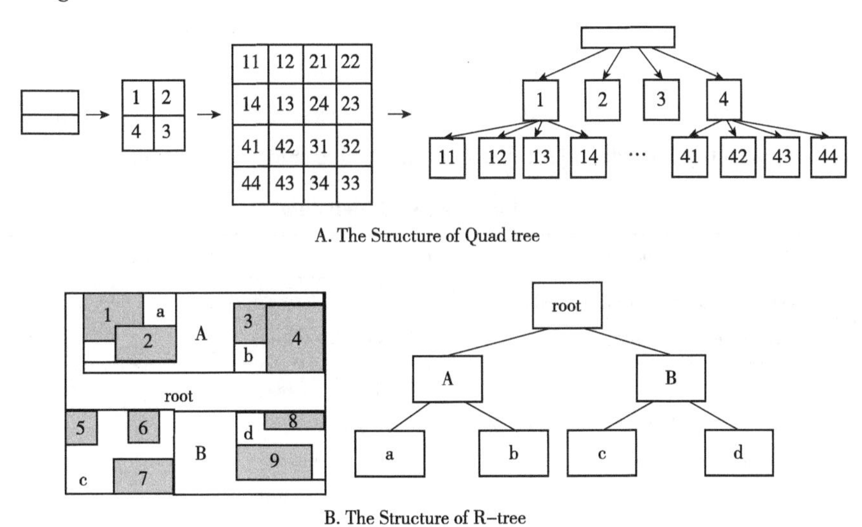

Figure 6-25 Example of Spatial Index

6.3.7 Metadata

Metadata (元数据) is critical for sharing tools, data, and maps and for searching to see if the resources you need have existed. Metadata describes GIS resources consistent with the way the library catalog describes books.

Metadata is data about data and data usage (元数据是指关于数据的数据及其使用的各个方面). It specifies the information about what it is about, where it is to be found, who needs to get it, how much it costs, in what format it is available, what is the quality of the data for a specified purpose, what spatial location does it cover and over what time period, when and where the data were collected and by whom and what purposes the data have been used for, by whom and what related data sets are available, etc.

As information has become increasingly digital, metadata are used to describe digital data by usingmetadata standards specific to a particular discipline. By describing the contents and context of data files, the quality of the original data/files is greatly increased.

Metadata management is the end-to-end process and governance framework for creating, controlling, enhancing, attributing, defining and managing a metadata schema, model or other structured aggregation (元数据管理是创建、控制、加强、归属、定义和管理元数据架构、模型或其他结构化的聚合的首尾相连的过程和管理体系).

The concept of metadata is ambiguous, as it is used for two fundamentally differenttypes. Although the expression "data about data" is often used, it does not apply to both in the same way. Structural metadata, the design and specification of data structures, cannot be about data, because at design time the application contains no data. In this case the correct description would be "data about the containers of data". Descriptive metadata, on the other hand, is about individual instances of application data; the data content (Table 6-4).

Metadata contains information of identification, data quality, data organization, spatial reference, entity and attribute, distribution, metadata reference, citation, time period, and contact. Therefore, the functions of metadata include resource discovery, organizing e-resources, facilitating interoperability, digital identification, archiving and preservation.

Table 6-4 Types of Metadata

Type	Definition	Examples
Administrative	Metadata used in managing and administering information resources	Acquisition information Location information
Descriptive	Metadata used to describe or identify information resources	Cataloging records Finding aids Specialized indexes Annotations by users
Preservation	Metadata related to the preservation management of information resources	Documentation of physical condition of resources Documentation of actions taken to preserve physical
Technical	Metadata related to how a system functions or metadata behave	Hardware and software documentation Digitization information-Tracking of system response times

Metadata collected for a particular type of information resource. Every element in the scheme is assigned a name and the meaning of the element is specified. Optionally the rules on how the element contents must be formulated and represented can be given and any controlled vocabularies to be used within the value can be specified.

Metadata standards are metadata schemes developed and maintained by someone, either a specific standard organization (such as ISO), or an organization that has taken on the responsibility for such an initiative.

Vocabulary

abstraction[æb'strækʃn] *n.* 抽象；抽象概念；心不在焉
access['ækses] *n.* 入口；通道 *vt.* 进入；存取
accuracy['ækjərəsi] *n.* 准确(性)；精确(性)
administrative[əd'mɪnɪstrətɪv] *adj.* 行政的；管理的
associated[ə'səʊʃieɪtɪd] *adj.* 联合的；相关的
algorithms['ælgərɪðəm] *n.* 算法
capture['kæptʃə(r)] *n.* 捕获；战利品 *vt.* 捕获；占领；夺取；吸引；留存
context['kɒntekst] *n.* 上下文；环境；背景
convenient[kən'viːnjənt] *adj.* 便利的；方便的
derivation[ˌderɪ'veɪʃn] *n.* 派生；推导；来历；衍生物；导数
discrepancies [dɪs'krepənsi] *n.* 差异；不一致；分歧
depict[dɪ'pɪkt] *vt.* 描述；描绘；画
emulsion [ɪ'mʌlʃn] *n.* 乳状液；感光乳剂
electromagnetic[ɪˌlektrəʊmæg'netɪk] *adj.* 电磁的
elimination[ɪˌlɪmɪ'neɪʃn] *n.* 除去；消除；淘汰
generation[ˌdʒenə'reɪʃn] *n.* 代；(产品类型的)代；产生；繁殖
hierarchical[ˌhaɪə'rɑːkɪkl] *adj.* 按等级划分的
intermediate[ˌɪntə'miːdiət] *adj.* 中间的；中级的 *n.* 调解人；媒介物 *vi.* 调解；干涉
integrity[in'tegriti] *n.* 完整性
merging[mɜːdʒɪŋ] *n.* 融合；归并；汇合；数据并合
minimum bounding rectangle(MBR) 最小外接矩形
overshoot[ˌəʊvə'ʃuːt] *v.* (射箭等)射过头；越过(目标)
photogrammetry [fəʊtə'græmətrɪ] *n.* 照相测量法
polar['pəʊlə(r)] *adj.* 两极的；对立的 *n.* 极线；极性
projection[prə'dʒekʃn] *n.* 投影；投射；规划；发射；凸出物；预测；放映
sheeting['ʃiːtɪŋ] *n.* 薄片；被单布
radiation[ˌreɪdi'eɪʃn] *n.* 辐射；放射线
stretch[stretʃ] *v.* 伸展；延伸；张开；夸大 *n.* 伸展；张开；弹性 *adj.* 可伸缩的
spectrum['spektrəm] *n.* 系列；幅度；范围；光谱；频谱
sliver['slɪvə(r)] *n.* 裂片；细长条；梳棉 *v.* (使)成小片；剖成长条
theodolite[θi'ɔdəlaɪt] *n.* 经纬仪
undershoot[ˌʌndə'ʃuːt] *v.* 飞机未达预定点着陆；脱靶

Questions for Further Study

1. What are the advantages of batch vectorization over manual table digitizing?
2. What quality assurance steps would you build into a data collection project designed to con-

struct a database of land parcels for tax assessment?

3. Why do so many geographic data formats exist? Which ones are most suitable for selling vector data?

4. Identify a geographic database with multiple layers and draw a diagram showing the tables and the relationships between them. Which are the primary keys, and which keys are used to join tables? Does the database have a good relational design?

5. What are the advantages and disadvantages of storing geographic data in a DBMS?

Chapter 7

Geographical Query and Measurements

Geographical query involves the activities of exploring the general characteristics and trends in the geospatial data, taking a close look at data subsets, and focusing on some spatial relationships between them. Query methods allow us to interact with geodatabases using pointing devices and keyboards. Effective geographical queries also require interactive and dynamically linked visual tools (地理查询包含探究地理空间数据总体特征和趋势、详查数据集及它们之间的空间关系等活动。查询方法使我们能够利用鼠标和键盘与地理数据库进行交互。有效的地理查询也需要具有交互性和动态性相关的可视化工具). Measurements have simple numerical values that describe aspects of geographical data. They are common and very useful in GIS.

7.1 Basic Concept

Query is the most basic spatial analysis operation, in which the GIS is used to find or calculate specific geographical data or geographical information. Geographic query is a kind of interactive method to explore spatial data in different linked views (Figure 7-1). It is also helpful for us to find *anomalies* (异常值) and *outliers* (极端值) in the geodatabase, or to find particular features in images. In a sense, geographic query can be referred to as data mining in large masses of spatial data (地理查询是在不同相关视图中探索空间数据的一种交互方法，它有助于我们在地理数据库中发现异常值和极端值，或在影像中找到特定的特征。某种程度上，地理查询是指对海量空间数据的数据挖掘).

In the process of geographic query, no changes occur in the geodatabase, and no new data are produced (地理查询过程中，地理数据库不发生变化，也不产生新数据). The geodatabase can be searched to present records that meet certain criteria. The operations vary from simple and well-defined queries like "how many houses are found within 1 km of this point", to complex and *vaguer* (模糊的) questions like "which is the closest city to Xi'an going north".

Queries may be operated by putting in or clicking fields and operations in a query box (查询

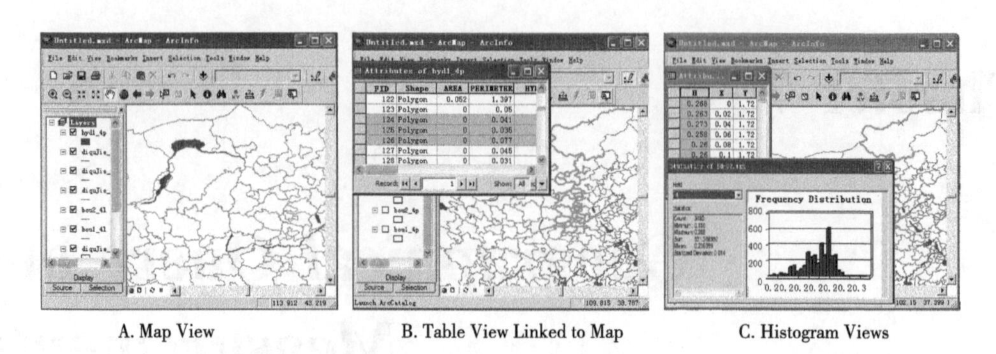

A. Map View B. Table View Linked to Map C. Histogram Views

Figure 7-1 Examples of Views

可以通过手工输入进行，也可以通过点击查询对话框中的字段和操作符来进行). We can display maps, graphs and tables in multiple but dynamically linked windows so that, when we select a data subset from a table, it automatically highlights the corresponding objects in a map. The importance of maps in GIS has added geographical visualization to the list of query activities. The purpose of geographical query is to better understand the geospatial data and to provide a starting point in formulating research questions and *hypotheses* (假设).

Reasoning is the process of forming interpretation, judgment or conclusion from some events, phenomena or behaviors (推理是从某些事件、现象或行为中形成解释、判断或结论的过程).

Query and reasoning focus on the process of reasoning from the results of a limited sample and making generalizations about a database (entire spatial data database). For instance, it allows us to determine whether a pattern of points could have arisen by chance, based on the information from a sample. Query and reasoning also allows the geospatial data to be viewed from different perspectives, making it easier for information processing and *synthesis* (综合). Query and reasoning is similar to exploratory data analysis in statistics, and used to explore spatial data structure and to focus on a data subset of interest.

7.2 Geographical Query

7.2.1 Simple Display and Query

Mapmaking is a *routine* (常规的) GIS operation. As effectively communicating spatial information, maps display the locations or all objects using points and arcs, with or without background. Normally, we can derive maps (or thematic maps) from geographical query. And we prepare maps for geographical visualization and presentation. A map for simple display usually has a number of elements: title, subtitle, body, legend, north arrow, scale bar, acknowledgment, neatline and border (用作简单显示的地图通常包含一些要素：标题、副标题、主体、图例、指北针、比例尺、制作单位、边线及边界线).

(1) Relational Query (关系查询)

Based on their spatial relationships of geographic features, relational query works with a relational geodatabase, which may consist of many separate but interrelated tables. Features to be selected may be in the same layer or, more commonly, in different layers. A query of a table in a relational geodatabase not only selects a data subset in the table, but also selects records related to the subset in other tables.

To use a relational geodatabase, we must design the special structure of the database. So that, different systems use different ways of formulating queries. The designation of keys is in relating tables, and a data dictionary listing and describing the fields in each table. Although the general concept is the same, the structure of these expressions varies from one system to another. Structured Query Language (SQL) is an attempt to provide a "standard" way in querying database. The user can select object of interest and produce maps by using relational query.

(2) Display Map by Using Simple Symbols

On a map, the real-world features are replaced by symbols in their corresponding spatial location at a reduced scale. Attributes and entity types can be displayed with color-coded line patterns and dot symbols. These elements work together to bring spatial information to map readers.

The map view of a dataset shows its contents in the visual form, and opens many possibilities for querying. When the user points to any location on the screen, the GIS should display the pointer's coordinates with the units appropriate to the dataset's projection and coordinate system. If the dataset is raster, the system might display a cell's row and column number, or its coordinate system if the raster is adequately georeferenced (tied to some earth coordinate system).

The main advantages of visual spatial query are: natural, intuitive, easy to operate, complex queries can be composed with different symbols. However, it also has some disadvantages. For example, when the space constraint conditions are complex, it is difficult to use symbols to describe; ambiguity may appear when two-dimensional symbols are used to represent the relationship between graphs. It's hard to express a "not" relationship; it is not easy to restrict the range (circle, rectangle, polygon, etc.); it is unable to screen location query, etc.

7.2.2 Spatial Query

Spatial query is the process of selecting features based on their geographical or spatial relationship to other features. A spatial query is a special type of database query supported by geodatabases. The queriy differs from SQL query in several important ways. Two of the most important are that it allows the use of geometry data types such as points, lines and polygons and considers the spatial relationship between these geometries. For instance, you might be interested in finding out what features are within a certain distance of other features, what are adjacent to other features, what are contained inside other features, or what intersects other features.

In addition, many types of spatial relationships can be discovered and identified by applying a series of spatial query operators to the geographical objects. We may select features using a *cur-*

sor (光标), a graphic, or the spatial relationship between features. Spatial query refers to the process of retrieving a data subset from a layer by working directly with features. The results can be simultaneously inspected in the map, linked to the highlighted records in the table, and displayed in charts. They can also be saved as a new data set for further processing.

The table view of a dataset shows a rectangular array, with the objects organized as the rows and the attributes as the columns. This allows the user to see the attributes associated with objects at a glance, in a convenient form. There will usually be a table associated with each type of object — points, lines, areas, background, or links.

Spatial query is a kind of commonly used spatial data query. According to the geometric positions of points, lines, rectangles, circles or other irregular objects, the ground objects of interest are selected to obtain the attributes, spatial positions, spatial distribution and spatial relations with other spatial objects of the query object. Point-by-point query conditions: given a mouse point, query to its nearest object and capture the attribute points; window query (according to rectangle, circle, polygon query) conditions are divided into window containing region and passing region (Figure 7-2).

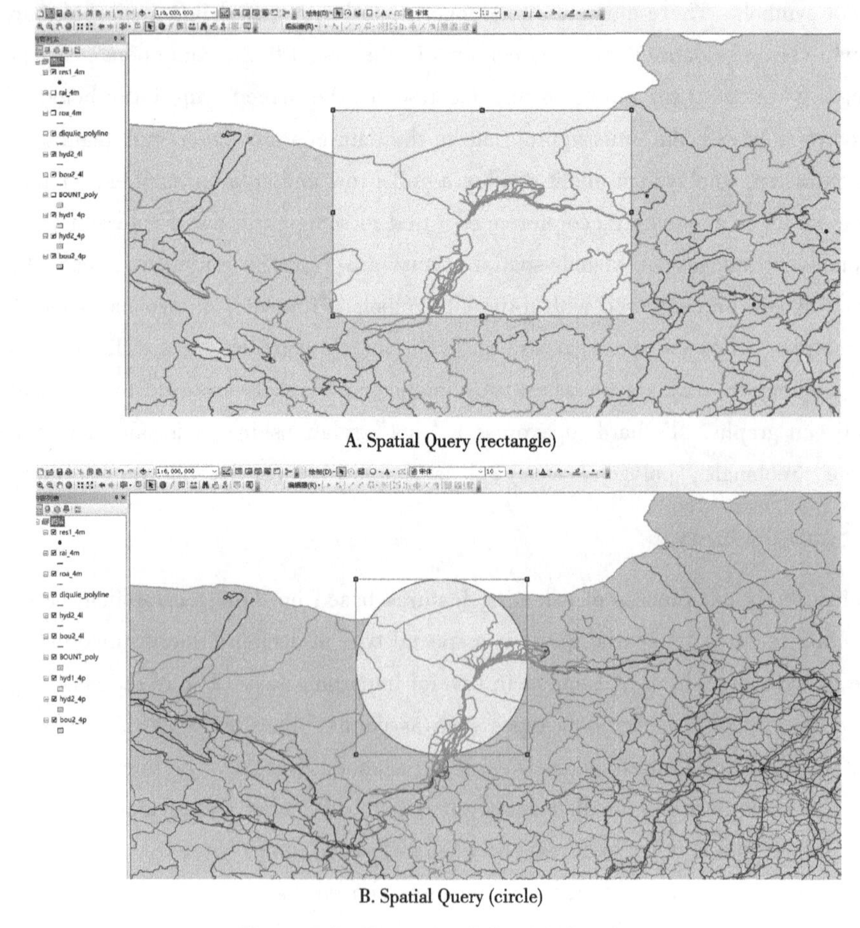

A. Spatial Query (rectangle)

B. Spatial Query (circle)

Figure 7-2 Example of Spatial Queries

The attribute query helps us select or view certain data on the map based on data's attributes. An attribute query creates a single query on one map layer. By linking a *scatterplot* (散布式绘图或散点图) (table view) with a map view, it is possible to select points in the scatterplot and see the corresponding objects highlighted on the map (Figure 7-1B). This kind of linkage is very useful in examining residuals. However, the query can have compound criteria. For instance, you can define a query on a province's layer, which selects provinces with a population greater than a value entered by the user and with a name that begins with a value entered by the user.

Additionally, the attribute query supports associated tables and stand-alone tables. Associated tables simply define a relationship between two tables without *appending* (附加的) the associated data in each table. In many cases, the relates point toward stand-alone tables, which lack geometry but often contain valuable attribute information. For example, if you published a map service that contained relates between city parcels and a stand-alone table of parcel owners, you could use the Attribute Query to display the owners and location of the parcel in the query results.

When you *configure* (安装/构建) the attribute query, you create an easy-to-understand form that will guide users through the process of making the query. Thus, users of your application will not have to know the details about the dataset, nor will they have to construct a SQL statement to query the data.

7.2.3 Spatial Relationship Query

When constructing a spatial relationship query, specify the type of spatial relationship to look for and the geometry to compare. Spatial relationship queries return true or false; geometries either participate in each other in a specific spatial relationship, or they do not.

Suppose both the map view and the table view are displayed on the screen simultaneously. Linkage allows users to select objects in one view, perhaps by pointing and clicking, and to see the selected objects highlighted in both views.

In most cases, you would use a spatial relationship query to filter a result set by placing it in the WHERE *clause* (语句).

For example, if you have a table that stores the locations of proposed development sites and another table that stores the locations of the archaeologically significant sites, you might want to make sure that the features in the development sites table do not intersect those in the archaeological sites. You could issue a query to make sure none of the development sites intersect archaeology sites and, if so, return the ID of those proposed developments.

(1) Adjacency Relation Query

The adjacency relation query can be point-to-point adjacency query, line-to-line adjacency query, or polygon-to-polygon adjacency query. Adjacency relation query can also be a query involving the linear and planar feature information adjacent to a node, such as finding the vacant land adjacent to the park or residential areas adjacent to the flood area.

(2) Inclusion Relation Query

The inclusion relation query can query a class of features contained in a surface feature or a surface feature that contains a certain feature. The included features can be point features, line features or plane features, such as the distribution of commercial outlets in an area.

(3) Association Relation Query

The association relation query is the query of topological relationship between different elements in the space, which can query the relevant information of the linear feature associated with a certain point feature, or the relevant information of the planar feature associated with the linear feature. For example, query the power line to which a given pole is connected.

7.2.4 Hypertext Query

Hypertext query treats graphics, images, characters, etc. as texts, and sets some "hot spots" which can be text, keys, etc. After clicking the "hot spot" with the mouse, you can pop up the explanatory information, play the sound, complete some work, etc. But hypertext queries can only be set up in advance, and users cannot build the various queries they require in real time.

7.2.5 Extended SQL Query

SQL is a powerful language used to define one or more criteria that can consist of attributes, operators, and calculations. It also is a standard language for querying tables and relational databases.

The standard language for database query adopted by mainstream databases is SQL (ISO Standard ISO/IEC 9075). SQL may be used directly via an interactive command line interface. When a query is specified for an update or search cursor, only the records satisfying that query are returned. It may be compiled in a general-purpose programming language (e. g. C/C++/C#, Java, or Visual Basic); or it may be *embedded* (嵌入) in a graphical user interface (GUI). A SQL query represents a subset of the single table queries that can be made against a table in a SQL database using the SQL SELECT statement. SQL is a set-based rather than a procedural (e. g. Visual Basic) or object-oriented (e. g. Java or C#) programming language designed to retrieve sets (row and column combinations) of data from tables.

In SQL, data definition language statements are used to create, alter, and delete relational database structures. The syntax used to specify the WHERE clause is the same as that of the underlying database holding the data. For example:

```
SELECT * FROM counties WHERE "NAME"= "Shaanxi"AND "POPULATION">= 20000//
在名为 counties 的文件中,查找 NAME 为 Shaanxi,且 POPULATION 大于等于 20000 的所有县
```

CREATE TABLE loess_plateau (lp_id INTEGER,name VARCHAR2(32),shape SDO_GEOMETRY);

//创建表 loess_plateau，该表包含整型的 lp_id 列、字符串型 name 列和 SDO_GEOMETRY shape 列

```
INSERT INTO loess_plateau (OID,address,city_code,loc)
VALUES (1,
'yangling',
712100,
st_point (0.00003,0.00051,null,null,0)
)
//在 loess_plateau 表中添加记录
```

7.3 Geographical Measurements

Measurements are simple numerical values that describe aspects of geographical data. They are common and very useful in GIS. They represent simple properties of objects, like length, area or shape, and the relationships between pairs of objects, like distance or direction.

Many GISs have measurement tools. For instance, to activate a measuring pointer, an *icon* (图标) is clicked to measure length and distance between pointed locations. The length of a single line is simply measured between the start and end points, complex line measures the length between points and gives the cumulative total length (same topology as measuring a chain in a line or polygon feature). Some GISs offer *sophisticated* (复杂的) spatial measurements that can calculate distances between many points automatically, rather than depending on individual manual operations.

7.3.1 Distance and Length

In planar rectangular coordinate system, *Pythagoras's Theorem* (勾股定理，毕达哥拉斯定理) and the straight-line distance are used for measuring distance and length between two points on a planar coordinate system. The square of *the length of the hypotenuse* (直角三角形的斜边) is equal to the sum of the squares of the lengths of the other two sides of a right triangle (Figure 7-3A). Furthermore, the length of a polyline tends to be shorter than the length of the object it represents (Figure 7-3B), and longer than its horizontal projection (Figure 7-3C). If a line is represented as a polyline, or a series of straight segments, then its total length is simply equal to the sum of the lengths of each segment, and each segment length can be calculated using the Pythagorean formula.

However, in GIS, the simple rules are not sufficiently accurate for measuring distances on

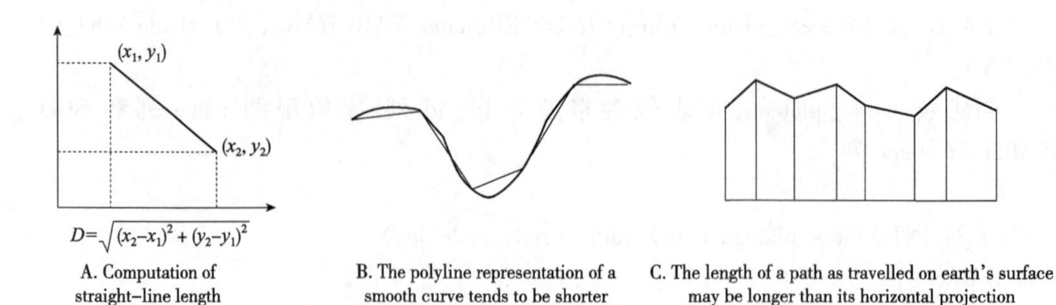

A. Computation of straight-line length

B. The polyline representation of a smooth curve tends to be shorter

C. The length of a path as travelled on earth's surface may be longer than its horizontal projection

Figure 7-3 Measurements of Distance and Length

the earth's surface properly. Therefore, the standard and simplest method for calculating the shortest distance between two points on the earth's surface may be the *Great Circle Distance* (*geodesic distance*-大圆距离，测地距) equation, and it is a *metric* (度量) for determining distance between points in space:

$$\cos(d) = \sin(a)\sin(b) + \cos(a)\cos(b)\cos(c) \tag{7-1}$$

Where, d—the angular distance between points A and B in degrees, a and b—the latitude of A and B, and c—the difference in longitude between A and B. Many GIS users are using spatial data in geographical coordinates directly for data display and even simple analysis. Distance and length from such spatial data are usually derived from the shortest *spherical distance* (球面距离) between points.

7.3.2 Area

Area(面积) is the size of a surface. For example, different shapes in Figure 7-4 all have the same area of 10.

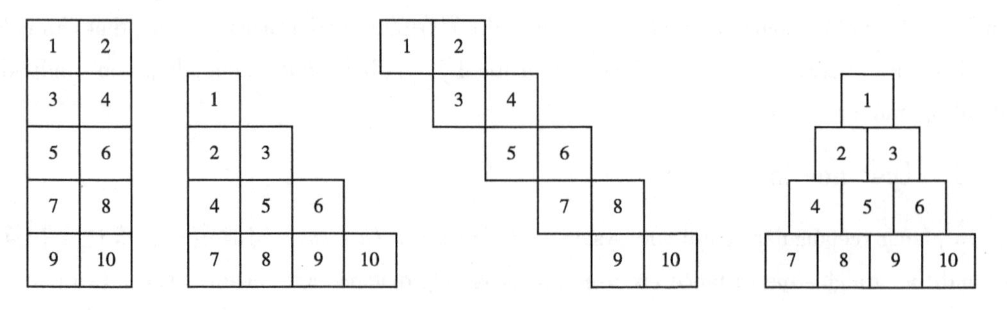

Figure 7-4 Different Shapes All Have the Same Area of 10

Areas of simple planar shapes such as Triangle, Square, Rectangle, *Parallelogram* (平行四边形), *Trapezoid* (梯形), Circle, *Ellipse* (椭圆), and *Sector* (扇形), can be measured by special formulas. For irregular complex polygons, we can break them up into two or more simpler shapes. Sometimes, if we know the coordinates of each corner point (vertex), we can work out the total area by calculating each area along each line segment to the x-axis.

7.3.3 Shape

The shape of an object located in some space is a geometrical description of the part of that space occupied by the object, as determined by its external boundary-abstracting from location and orientation in space, size, and other properties such as color, content, and material composition.

Simple shapes can be described by basic geometrical objects such as a set of two or more points, a line, a curve, a plane, a plane figure (e.g. square or circle), or a solid figure (e.g. cube or sphere). Most shapes occurring in the physical world are complex. Some, such as plant structures and coastlines, may be so *arbitrary* (任意的) as to *defy* (挑衅) traditional mathematical description-in which case they may be analyzed by differential geometry, or as *fractals* (分形).

There are numbers of ways to describe and measure a complex shape. However, three commonly used shape factors-circularity, convexity and elongation are required.

(1) Circularity

One measure of shape is to quantify the "closeness" to a perfect circle. For this, the factor of *circularity ratio* (圆度比) is introduced and defined as: Circularity ratio is the ratio of the area of the test shape to the area of a circle (the most compact shape) having the same perimeter as the text shape (area-perimeter ratio).

$$\text{Circularity}=\frac{A}{A_{\text{circle}}}=\frac{A}{\pi\times\left(\frac{P}{2\pi}\right)^2}=\frac{4\pi\times A}{P^2} \tag{7-2}$$

Where A is the area of test shape, and P is its perimeter. The circularity ratio C will be greater than 0 and less than or equal to 1. For a long, thin shape it is 0; for a square, it is $\pi/4$; and for a circle, it is 1.

(2) Convexity

Convexity (凸性，凸曲度) is the ratio of the "*convex hull perimeter* (外接凸多边形周长，perimeter of the convex outline of the object)" to the actual perimeter of the object.

$$\text{Convexity}=\frac{P_{\text{convex hull}}}{P} \tag{7-3}$$

Convexity also has values in the range 0−1. A smooth shape has a convexity of 1 as the convex hull perimeter is exactly the same as the actual perimeter. A very *spiky* (尖钉形的) or irregular object has a convexity closer to 0 as the actual perimeter is greater than the convex hull perimeter due to the fine surface features.

(3) Elongation

Elongation (延伸性) is defined as [1−Aspect Ratio (高宽比，纵横比)] or (1−Width/Length).

$$\text{Elongation}=1-\frac{\text{Width}}{\text{Length}} \tag{7-4}$$

As the name suggests, it is a measure of elongation and again has values in the range 0–1. A shape *symmetrical* (对称的) in all axes such as a circle or square will have an elongation value of 0 whereas shapes with large aspect ratios will have an elongation closer to 1 (Table 7-1).

Table 7-1 Illustration of Shape Descriptors

Shape	Circularity	Convexity	Elongation
	1	1	0
	0.47	1	0.82
	0.89	1	0
	0.52	1	0.79
	0.47	0.70	0.24
	0.21	0.73	0.83

7.3.4 Continents or Islands

Whether an area is a continent or an island can be determined by analyzing the number of enclosed islands and zones that make up the region itself. If the number of zones is greater than the number of enclosed islands, then the region is closer to the shape of "island". Otherwise, when the number of enclosed islands is greater than the number of zones, then the region is closer to the shape of "continent" (Figure 7-5).

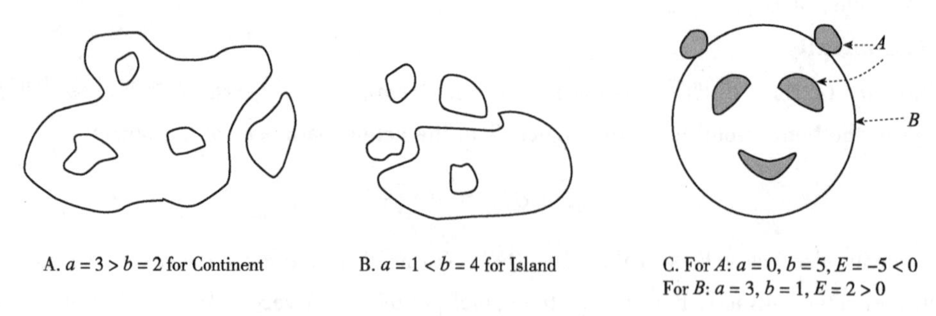

Figure 7-5 Judgment of Continent or Island

(Note: a=number of enclosed islands; b=number of zones which make up the region itself; E=Euler number)

In addition, Whether an area is closer to a continent or an island can be determined by it Euler number. The Euler number is calculated by subtracting the number of islands (zones) which make up the region itself from the number of enclosed islands in the region. If the Euler number is greater than 0, the region is closer to a "continent". Otherwise, if the Euler number is less than 0, then the region is closer to "islands".

Vocabulary

appropriate[ə'proʊpriət] *adj.* 适当的；相称的 *vt.* 占用；拨出(款项)

adequately['ædɪkwətli] *adv.* 足够地；充分地；适当地

adjacency[ə'dʒeɪsənsɪ] *n.* 毗邻；四周；邻接物

ambiguity[ˌæmbɪ'gjuːəti] *n.* 含糊；不明确；暧昧；模棱两可的话

at a glance 一看就；看一眼便

compile[kəm'paɪl] *vt.* 编译；编制；编纂

circularity[ˌsɜːkjə'lærəti] *n.* 循环性；圆；环状

continent['kɒntɪnənt] *n.* 大陆；洲；(the Continent)欧洲大陆

convexity[ˌkɒn'veksətɪ] *n.* 凸状；凸面；凸性

coordinate[koʊ'ɔːrdɪneɪt] *v.* (使)协调；(使)一致；(使)同等 *n.* 坐标；同等的人物；配套服装 *adj.* 同等的；等位的；(大学)男女分院制的

criteria[kraɪ'tɪəriə] *n.* 标准；尺度；准则

data mining 数据挖掘技术

dynamically[daɪ'næmɪkli] *adv.* 动力地；动态地；有活力地

elongation[ˌiːlɒŋ'geɪʃn] *n.* 延伸

enclose[ɪn'kloʊz] *vt.* 圈起；围住；附上；封入

extended[ɪk'stendɪd] *adj.* 长期的；广大的；伸展的动词 extend 的过去式及过去分词形式

equation[ɪ'kweɪʒn] *n.* 相等；均衡；方程式；等式

Euler number 欧拉数

geodatabase 地理数据库

geodesic[ˌdʒiːoʊ'desɪk] *adj.* 测地学的；测量的；测地线的 *n.* 测地线

highlighted['haɪlaɪtɪd] *adj.* 突出的动词 highlight 的过去式和过去分词

hypertext ['haɪpətekst] *n.* [计] 超文本(含有指向其它文本文件链接的文本)

island['aɪlənd] *n.* 岛；岛屿；岛状物

linkage['lɪŋkɪdʒ] *n.* 联系；连合；连锁；结合

metric['metrɪk] *adj.* 公制的；米制的；十进制的 *n.* 标准；度量

neatline['niːtlaɪn] *n.* 准线(图表边线；墙面交接线)

outlet['aʊtlet] *n.* 出口，排放孔；[电] 电源插座；销路；发泄的方法；批发商店

parallelogram[ˌpærə'leləgræm] *n.* 平行四边形

planar['pleɪnər] *adj.* 平面的；二维的；平坦的

purpose['pɜːrpəs] *n.* 目的；决心；意图；议题 *v.* 打算；决意

Pythagorean *adj.* 毕达哥拉斯哲学 *n.* 毕达哥拉斯哲学的

region['riːdʒən] *n.* 地区，区域；行政区；(首都以外的)地方(the regions)；身体部位；领域，界

residual[rɪ'zɪdʒuəl] *adj.* 剩余的；残余的 *n.* 剩余部分；(复)追加酬金

resort[rɪ'zɔːrt] *n.* (度假)胜地；手段；凭借 *vi.* 诉诸；常去

retrieving[rɪ'triːv] *vt.* 恢复；挽回；取回
scatterplot 散布式绘图法
segment['segmənt] *n.* 部分；弓形；瓣；段；节 *vt.* 分割
solid['sɒlɪd] *n.* 固体；立体图形
subset['sʌbset] *n.* 子集
subtitle['sʌbtaɪtl] *n.* 副题(书本中的)；说明或对白的字幕 *vt.* 给…加副标题；给…加字幕
symbol['sɪmbl] *n.* 符号；象征；标志
suppose[sə'poʊz] *vt.* 假设；假定；认为；想；应该；让(虚拟语气) *vi.* 推测
sufficiently[sə'fɪʃntli] *adv.* 足够地；充分地
visual['vɪʒuəl] *adj.* 视觉的；视力的；看得见的；形象的 *n.* 画面；图像
visualization[ˌvɪʒuəlaɪ'zeɪʃn] *n.* 可视化；形象化

Questions for Further Study

1. The result of geographic query can be visualized in different linked views, please demonstrate these different views as possible as you can according to your experiments in the lab.

2. How to query a destination in a Geodabase? Please list all methods you may use.

3. Sum up general area formulas of different regular shape.

Chapter 8

Spatial Analysis

What distinguishes GIS from other types of information systems is its spatial analysis functions (区分 GIS 和其他类型信息系统的是其空间分析功能). These functions use geographic data in the GIS database to answer questions about the real world. In a sense, spatial analysis is the crux of GIS because it includes the transformations, manipulations, and methods that can be applied to geographic data to add value, to support decisions, to reveal patterns and *anomalies* (异常). In other words, spatial analysis is the process in which we turn raw data into useful information, in pursuit of scientific discovery, or more effective decision making.

8.1 Introduction

The advantage a GIS can provide is the capability for transforming the original spatial data to answer users' questions. Such transformations are often referred to as "data analysis" capabilities. However, most so-called "analysis" capabilities of today's GIS are in fact data manipulation and maintenance, very rare of which can actually tell us something by "analyzing" spatial data.

Analysis is the process of identifying a question or issue to be addressed, modeling the issue, studying the model results, interpreting the results, and possibly making a recommendation (分析是确定一个要解决的疑问或问题，模拟该问题，研究模拟结果，解释结果及可能提出的建议的过程).

Analysis refers specifically to the transformation of parsed data (分析特指可解析的数据转换). "Analysis" is the process to resolve and separate the referencesystem into its parts to illuminate their nature and interrelationships, and to determine general principles of behavior (分析是融合与分离参考系用以阐明其自然属性和相互关系，并确定总体行为规则的过程).

Spatial analysis is an analytical technique associated with the study of locations of geographic phenomena together with their spatial dimensions and their associated attributes (空间分析是一个针对地理现象的区位及其空间维数和相关属性进行研究的分析技术). Spatial analysis is useful for *evaluating suitability* (适应性评估), estimating and predicting, and interpreting and understanding the location and *distribution* (分布) of geographic features and phenomena.

Spatial analysis can reveal things that might otherwise be invisible-it can make what is explicit or implicit. Some methods of spatial analysis were developed longbefore the advent of GIS, and carried out by hand, or by measuring devices like the ruler. The term *analytical cartography* (分析制图学) sometimes refers to the methods of analysis that can be applied to maps to make them more useful and informative, and spatial analysis using GIS is its logical successor in many ways.

Effective spatial analysis requires an intelligent user rather than a powerful computer. A large body of methods of spatial analysis have been developed over the past century or so, and some methods are so highly mathematical that it sometimes seems that mathematical complexity is an indicator of the importance of a technique (过去的约一个世纪以来，涌现出大量的空间分析方法，一些方法很大程度上是建立在数学基础上的，它们对数学的依赖程度如此之高，以至于有时候数学的复杂性似乎成了某项技术重要性的标志). But the human eyes and brains are also very sophisticated processors of geographic data and excellent detectors of patterns and anomalies in maps and images. So the approach taken here is to regard spatial analysis as spread out along a continuum of sophistication, ranging from the simplest types that occur very quickly and intuitively when the eyes and brains look at a map, to the types that require complex software and sophisticated mathematical understanding. Spatial analysis is best seen as a *collaboration* (合作) between the computer and the human, in which both play vital roles. Spatial analysis helps us in situations when our eyes might otherwise deceive us.

There are many possible ways of defining spatialanalysis. However, the basic idea that information on locations is essential, and analysis without knowledge of locations is not spatial analysis (不包括位置信息的分析并不是空间分析): as a set of methods, results from spatial analysis are not necessarily invariant under changes in the locations of the objects being analyzed (空间分析是一系列的方法，随着研究对象位置的改变其结果并非保持不变的).

Geographical analysis allows the study of real-world processes by developing and applying models. A GIS enhances this process by providing tools which can be combined in a meaningful sequence to develop new models. These models may reveal new or previously unidentified relationships, thus increasing our understanding of the real world. Results of geographical data analysis can be communicated with maps, reports, or both.

Spatial analysis is the vital part of GIS. It can be done in two ways: vector-based analysis and raster-based analysis.

8.2 Vector-based Analysis Capabilities

8.2.1 Attribute Data Operations

Many operations can be conducted on the attribute databases (the data tables). These can be divided into 4 categories: queries (or logical), arithmetic operations, statistical operations and reclassification . Attribute data is generally included in relational tables and it enhances our analytical capabilities through the spatial join function which *straddles* (跨越) attribute and spatial analysis.

(1) Queries = Select by Attributes

Queries include both comparison (=, >, <, >=, <=, <>) and Boolean (AND, OR and NOT) operators. These operators are used to perform queries.

(2) Arithmetic Operations

Arithmetic operators such as +, -, /, *, n (raised to the power of), √, Sin, Cos, Tan, etc. perform simple mathematical functions on values in the attribute database. Values calculated with these operators can be placed in a new field. i. e.

①Convert square meters (m^2) to hectares (hm^2) [e. g. divide by 10000]; results would be placed in a new field in the table.

②Convert driving distance to driving time [e. g. divide by average driving speed]; results would be placed in a new field in the table.

③Determine total volume (m^3) [e. g. multiply area (hm^2) by inventory volume (m^3/hm^2)]; results would be placed in a new field in the table.

(3) Statistical Operations

Statistical operations can also be performed on the attribute data. There are 2 options available: "statistics" and "summary".

"Statistics" provide a temporary pop-up table with the typical parameters: count, minimum, maximum, sum, mean and standard deviation. Plus the data are plotted in a histogram (frequency distribution).

"Summary" creates an output data table. Statistics are based on unique values in a chosen field. Selected fields from these operations are placed in a summary table. In the example of Table 8-1 and Table 8-2, the field "Group" was chosen and values on count, sum and mean were calculated.

(4) Reclassification

Reclassification is another operation conducted on attribute data. It results in a generalization

Table 8-1 Data Table

Group	Value	Group	Value
A	100	B	80
A	300	B	70
C	50	A	300

Table 8-2 Summary Table

Group	Count Value	Sum Value	Mean Value
A	3	700	233
B	2	150	75
C	1	50	50

(i. e. a simplification) of the original data set. For instance, raw property values in a data set can be put in 3 classes: Lower, Middle, and Upper. We typically use the legend editor (in ArcGIS the "Symbology" tab of the Layer Properties dialog box) to classify the data-altering the legend is temporary and we can change the coloring at any time (if we want a "permanent reclassification", e. g. a new map, then we would use Merge/Dissolve).

(5) Table Relations

Relating tables to each other involves joining or linking records between two tables. This may not be considered "an operation", but it allows us to relate outside source data to our themes and allows the features in our themes to be analyzed based on outside data.

(6) Spatial Join

As with relating tables, a spatial join will relate records between two tables. But the records are not joined based on a common attribute value (usually ID); instead records are joined based on "common location" (as defined by the coordinates of the spatial features). This type of operation is a combination of spatial and attribute; it is described in more details in section 8. 2. 2 below.

8. 2. 2 Spatial/ Geometric Operations

The spatial characteristics of maps (points, lines, polygons) can also be analyzed. Location, size and shape of the map, are defined by their coordinates with these operations. Spatial operations can be categorized as follows.

(1) Spatial Query-Select by Location

This is where features in one theme are selected based on their spatial relation (connectivity, containment, intersection, or nearness) to features in a second theme (i. e. select forest stands that contain an eagle's nest); new data is not created, just a set of features are selected. A few examples:

①Intersect-share geographic space (roads that cross creeks).

② Within a distance of-as the name implies, select features within the "buffer area" (wildlife trees within 20 m of river).

③Contain-feature has to be within (e. g. select forest stands that contain wildlife trees).

(2) Spatial Calculations

Simple spatial calculations determine areas, perimeters, and distances based on the coordinates (in ArcView these are accessed through the "shape" field as it contains the vertices); the calculations utilize the coordinates that define the features, but the results are stored in the database table (so this operation also straddles both attribute and geometric).

(3) Spatial Join

As previously stated, this operation is a mix of spatial operation and attributes operation. The end result is a join of two database tables, but the basis for the join is "coincident space". As with a "regular join" the relation has to be one-to-one or many-to-one between records in the "destination-to-source tables". For example, we could have two themes: "Cities and Countries" of the world. A spatial join could be done for "Cities" as the destination theme, as it yields a many-to-one relation (many cities to one country). A spatial join would thus bring data from the "Countries" theme to the "Cities" theme. However, a spatial join could not be done with "Countries" as the destination table as the relation would be one-to-many (Figure 8-1).

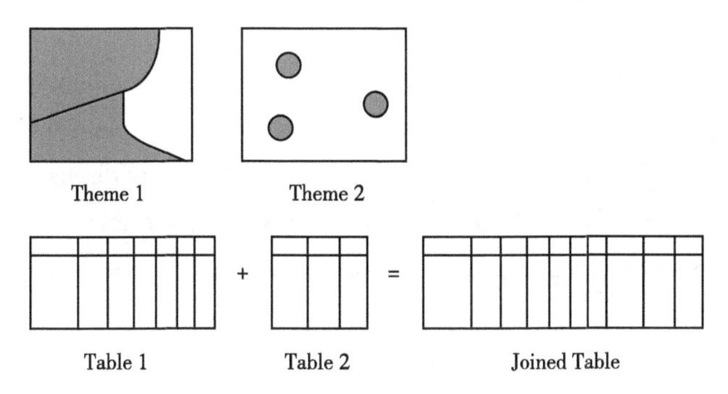

Figure 8-1 Spatial Join Operation Creates a Joined Table in Which Fields from Theme 2's Attribute Table are Appended to Theme 1's Attribute Table Based on the Relative Locations of the Features in the Two Layers (Themes)

(4) Topological Overlay

Overlay-bringing together two themes-the line work of the two themes are combined [lines are broken, new nodes and links/arcs are recognized, topology is redone, note that *sliver polygons* (多边形) may have to be eliminated] and the fields from both theme databases are combined into one new database.

Map overlay(叠置分析) is a common method of extracting hidden spatial information in GIS. It is to overlay each related data layer to form a new data layer and new spatial relation. This new data layer integrates attribute of old feature layers and it is assigned to new attribute. What should be noted is that feature layers that are overlaid must be based on the same coordinate system and *datum*(基准面).

In principle, map overlay is analysis of new feature attribute with some mathematical models. *Intersect* (逻辑并), *union* (逻辑交) and *subtract* (逻辑差) are involved in map overlay. According to different features, map overlay contains overlay of point and polygon, line and polygon, polygon and polygon. According to different operations, map overlay consists of *erase* (删除), identity, intersect, symmetrical difference, union and update.

Overlay is the process of superimposing two or more maps, through registration to a common co-ordinate system, so that the resultant maps contain the data from original maps for selected features (叠置是将两幅或多幅地图配准到一个共同的坐标系统中并叠加的过程，新产生的地图包含着来自原始地图中所选要素的数据). For route events, two event tables are overlaid to create an output event table that represents the union or intersection of the input. Although the term "overlay" can be applied to paper based maps, more often it applies to the use of digital data, nevertheless, the principle is the same.

①Suppose individual layers have planar enforcement. When two layers are combined (overlaid or superimposed), the result must have planar enforcement as well.

②New intersection must be calculated and created wherever two lines cross and a line across an area object will create two new area objects.

③When topological overlay occurs, spatial relationships between objects area updated for the new, combined map.

In practice, the first consideration for overlay is feature type. According to the graphical characteristic of the input layer, there are three common overlay operations: point-in-polygon, line-in-polygon, and polygon-on-polygon. To distinguish the layers in the following discussion, the layer that may be a point, line, or polygon layer is called the input layer, and the layer that is a polygon layer is called the overlay layer. In a point-in-polygon overlay operation, the same point features in the input layer are included in the output but each point is assigned with attributes of the polygon within which it falls. In a line-in-polygon overlay operation, the output contains the same line features as in the input layer but each line feature is dissected by the polygon boundaries on the overlay layer. Thus the output has more line segments than the input layer does. Each line segment on the output combines attributes from the input layer and the underlying polygon. The most common overlay operation is polygon-on-polygon, involving two polygon layers. The output combines the polygon boundaries from the input and overlay layers to create a new set of polygons. Each new polygon carries attributes from both layers, and these attributes differ from those of adjacent polygons.

According to the operation form, seven common types oftopological overlays include: Append, Union, Identity, Intersect, Update, Clip and Erase. Each of the overlay operations results in new datasets. For output datasets, feature geometry is almost always modified, and it is possible to update geometry attributes (e. g. area, perimeter, length). The recalculation of geometry, plus the joining of user attributes, is what gives the power to the overlay operations.

①Append: Appending is used to merge together multiple datasets that represent the same thematic data and have the same attributes, but are contiguous. For example, Append can put together a layer from many input layers. The input datasets can be feature classes of point, line or polygon, tables, rasters and raster catalogs.

②Unite: Uniting is a complete merging of two themes. The new output theme is composed of the entire map area of both themes and all the fields from both theme data tables. The order of inputs does not matter.

③Identify: Indentifying maintains all features of the input layer, but takes features from the identity layer that overlap with the input layer. The output layer's coordinate properties are dependent on which of the inputs is the identity layer. This is very similar to the Uniting, but it includes a clip to the polygon boundary of the input layer. The spatial *extent* (范围) of the output of identity operations using the same source data sets will differ if the order of the input and identity layers is switched, therefore, in the identity function, order of *precedence* (优先级) is important.

④Intersect: Intersecting is a merging of two themes but only areas which are common between the two inputs are included in the output and the attribute database is composed of all the fields from both theme data tables. The order of input layer and intersect layer does not matter.

⑤Update: Updating features from the "update layer" descend upon the input theme and replace whatever was underneath. In other words, Update replaces overlapping parts of the input layer with features from the update layer. An example would be updating a timber type map (input layer) with a cut block (update layer) where the cut block shape *supersedes* (取代) the timber types it overlaps. Order of precedence is important with Update.

⑥Clip: Clipping is *akin* (近似的) to pressing a cookie *cutter* (切割刀) onto a theme such that the new theme is a *miniature* (小型的) version of the first, the map area is defined by the overlay (cookie cutter) theme and the database comes only from the input (cookie dough) theme. It clips out parts of the input layer with the outer shape of the clip layer. Only attributes from the input layer are retained. For example, for cutting a map acquired elsewhere to fit a study area. The input may be a point, line, or polygon layer, but the clip layer must be a polygon layer.

⑦Erase: the polygons from the "erase layer" descend upon the input theme and eliminate that area. "Erase" removes parts of the input layer based on the spatial properties of the erase layer. Attributes from the input layer are passed to the output layer, and none of the erase layer's attributes are transferred to the output. As with some of the other topological overlay operations, the order of input and erase layer matters. For instance, if land were *expropriated* (征收) from a *woodlot* (林地) owner to create a park. The "park area" would be erased from the woodlot area.

Figure 8-2 and Table 8-3 illustrate the topological overlay geoprocessing operations.

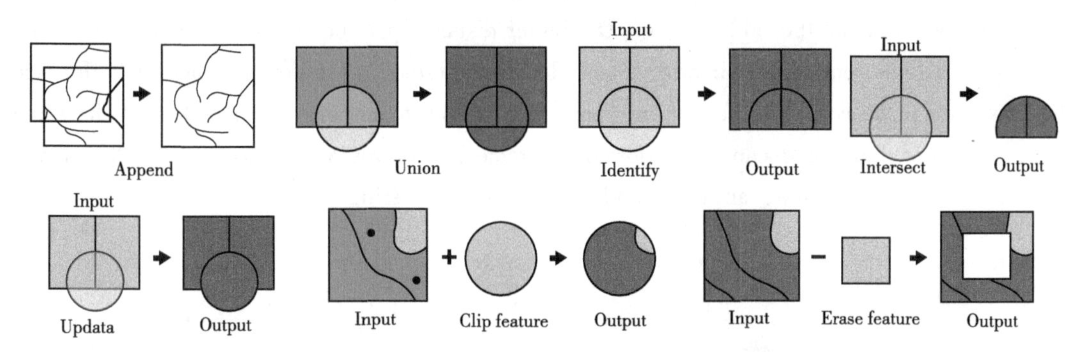

Figure 8-2 Seven Types of Topological Overlays

Table 8-3 Topological Overlay Geoprocessing

Input layer	Union layer	output layer	Input layer	Intersect layer	output layer
a b c d	2 1		2 1	a b c d	
Input layer	Identity layer	output layer	Input layer	Update layer	output layer
a b c d	2 1		2 1	a b c d	

（续）

Input layer	Union layer	output layer	Input layer	Intersect layer	output layer
2 1	a b c d	1 2 2 1 a b b b 1 2 2 1 c c d d 2 1	a b c d	2 1	a b c d
Input layer	Clip layer	output layer	Input layer	Erase layer	output layer
a b c d	2 1	a b c d	a b c d	2 1	a b c d
2 1	a b c d	1 2 1	2 1	a b c d	1 2 2 1

(5) Buffer

Buffer —Returns a geometry from which all the points are less than or equal to a user-defined distance（缓冲区——重现一个几何图形，在这个几何图形所有的点到中心的距离小于或等于用户自定义的距离）. Buffering creates a new theme with new polygon features (geometric objects) based on a specific measure of features in a source theme; buffers around points are circles, buffers around lines are "corridors" (snake-like with rounded ends) and bufffers around polygons are "donuts". Regardless of its variations, buffering uses distance measurements from select features to create the buffer zones. The buffer distance parameter can be entered as a fixed value or as a field containing numeric values. In other words, the buffer distance or buffer size does not have to be constant, it can vary according to the values of a given field. At the same time, an optional dissolve can be performed to remove overlapping buffers. Figure 8-3 shows instances of a point, a line, an area, and the results of buffering.

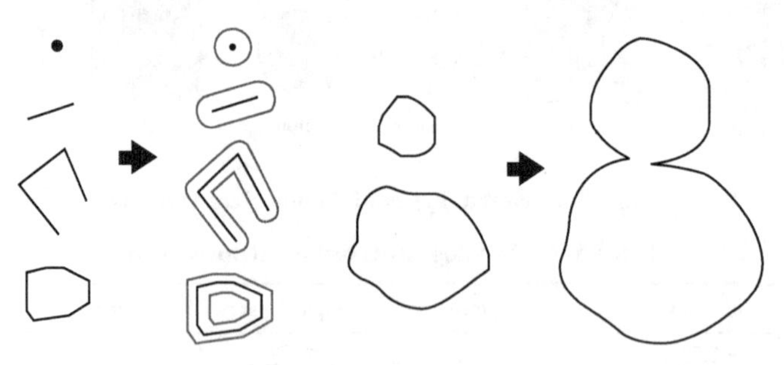

Figure 8-3　Buffering

Buffering around line features does not have to be on both sides of the lines, it can be on either the left side or the right side of the line feature. Likewise, buffer zones around polygons can be extended either outward or inward from the polygon boundaries. A feature may have more than one buffer zone. For example, a nature reserve may be buffered with distances of 5, 10, 15, and 20 kilometers, thus forming multiple rings around it.

Buffering algorithm creates a new area enclosing the buffered object. The application of these buffering algorithms fundamentally addresses the creation of zones around the target. e. g. protected zones around lakes, *reservoirs* (水库) or streams; zones of noise pollution around highways or airports; service zones around bus route; groundwater pollution zones around waste site, protection zones of ecological areas, etc.

Buffering is one of the most useful transformations in a GIS and is possible in both raster and vector formats. In the raster case, the result is the classification of cells according to whether they lay inside or outside the buffer, while the result in the vector case is a new set of objects. But there is an additional possibility in the raster case that makes buffering more useful in some situations.

(6) Dissolve

Dissolving is the process of removing boundaries between adjacentpolygons that have the same values for a specified attribute(融合是消除具有相同特定属性值的相邻多边形之间公共边界的过程)(Figure 8-4). Merging is to combine items from two or more similarly ordered sets into one set arranged in the same order (合并是将来自两个或多个相似的有序集合组成一个以同样序列组织的集合).

With dissolving, also known as merged polygons, boundaries between adjacent polygons with the same attribute values (i. e. class = poor) are "dissolved" and the two (or more) polygons are merged into one larger polygon; a new map layer (theme) results from the generalized data. This is the "spatial equivalent" to reclassification of attribute data.

Frequently data manipulation in working with area objects to *aggregate* (集群) areas is based on attributes. Commonly, they are used in a three-step procedure:

①Reclassify areas by a single attribute or some combination.

②Dissolve boundaries between areas of the same type by deleting the arc between two polygons if the relevant attributes are the same in both polygons.

③Merge polygons into large objects by recording the sequence of line segments that connect to form the boundary and assigning new ID numbers to each new object.

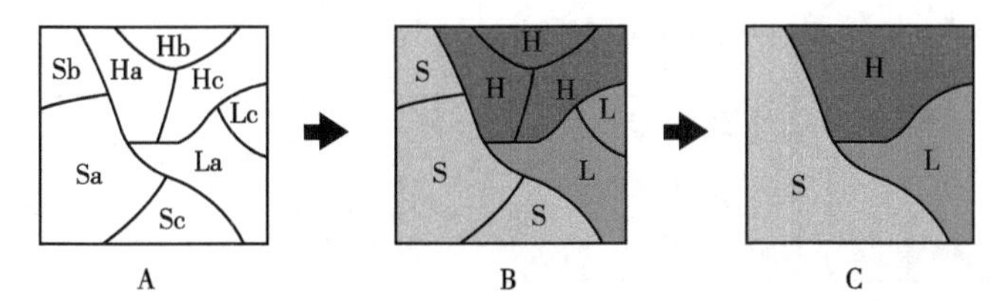

A. Classification map –Soil types (H, L, S) with population density (a>350, b>100, and c≤100 person/km^2);
B. Reclassify according to the attribute of Soil types; C. Result of dissolve and merge.

Figure 8-4 Data Manipulation Process of Dissolve

Note: Classification is a method of generalisation(分类是归纳的一种方法). In the process of classification, an attempt is made to group data into classes according to some common characteristics, thereby reducing the number of data elements. Classification tends to be based upon the attributes or characteristics of data rather than their geometry(分类倾向于基于数据的属性或特征而非其几何特性). In digital image processing, images are usually classified according to the spectral properties of the pixels composing the image. In spatial analysis, a map can be classified according to any attribute value, for example, soil types, population density, unemployment, etc. The result of performing classification is a thematic derived map.

(7) Network Routing

Network Routing-as the name implies, this type of analysis assesses movement through a network. Consider the difference between the shortest route and the fastest route. During the middle of the night the "shortest route" is likely the "fastest route". However, during rush hour we would consider traffic and use the "fastest route" (which may have a longer distance). The network routing is modeled using lines (arcs) and intersections (nodes). Arc-node topology provides information regarding connectivity. The attribute database would provide additional information regarding impedance to flow (or movement). Examples would include speed limit and traffic loads at different times of the day. There is a "cost" to making turns at intersections, i. e. you have to slow down rather than use just two wheels to make the turn. One-way streets would provide for an absolute barrier. As well, making a turn off an overpass onto a highway below would be *prohibited* (禁止). Other routings include the most efficient route (for making several stops or deliveries) and location-allocation (where school catchment areas can be determined based on road network and not just a straight-line distance).

(8) Pattern Analysis

Pattern analysis is the study of the spatial arrangements of point or polygon features in two dimensional space. Pattern analysis uses distance measurements as inputs and statistics (spatial statistics) for describing the distribution pattern. Spatial distributions are random, dispersed, and clustered (Figure 8-5). The spatial pattern measurement methods of point feature include quadrat analysis, nearest neighbor analysis, Ripley's K-function, etc. In addition to spatial location, the variation of an attribute at the location is also necessarily considered, a pattern analysis can detect if a distribution pattern contains local clusters of high or low values because pattern analysis can be a precursor to more formal and structured data analysis.

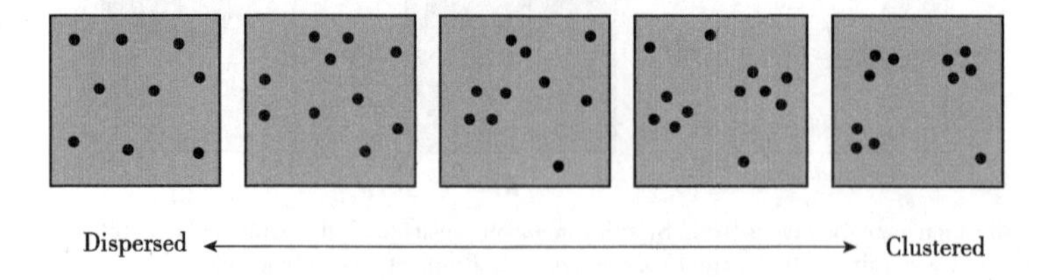

Figure 8-5 Patterns of Spatial Distribution

When analyzing the spatial distribution pattern, it actually measurs the spatial autocorrelation or spatial dependence. Spatial autocorrelation(空间自相关) measures the relationship among values of a variable according to the spatial arrangement of the values. The relationship may be described as highly correlated if values are spatially close to each other, and may be described as independent or random if no pattern can be discerned from the arrangement of values.

Spatial autocorrelation includes global spatial autocorrelation(全局空间自相关) and local spatial autocorrelation (局部空间自相关). Global spatial autocorrelation measures the correlation between values and spatial positions in the whole region. Popular measures are Morlan's *I*, Geary's *C* and G-Statistic. In fact, there are also local variations in spatial autocorrelation. For example, although the global indicator shows negative spatial correlation, there will be local aggregation patterns with positive spatial correlation. Local spatial autocorrelation measures each spatial feature. Popular measures of local spatial autocorrelation are local version of Morlan's *I*, Geary's *C* and G-Statistic, which is also known as the local indicators of spatial correlation (LISA). Local version of Moran's *I* calculates for each feature (point or polygon) an index value and a *Z* score. A high positive *Z* score suggests that the feature is adjacent to features of similar values, either above the mean or below the mean. A highly negative *Z* score indicates that the feature is adjacent to features of dissimilar values. Moran's *I*, either general or local, can only detect the presence of the clustering of similar values. It cannot tell whether the clustering is made of high values or low values. This has led to the use of the G-statistic, which can separate clusters of high values from clusters of low values.

8.3 Raster-based Analysis Capabilities

The following provides an overview of the various analytical functions available for raster data sets.

8.3.1 Analysis Environment

Geoprocessing analysis environment includes the workspace into which results will be placed, the extent, cell size, and coordinate system. It is important to be sure to set these parameters for the output layer prior to doing the analysis. Under general conditions, GIS packages require that all input raster layers should have the same resolution (cell size) and extents, although some of the more recent GIS softwares allow for layers with a variety of cell size and extents to be analyzed. In addition to this, a fifth parameter to consider is that of an analysis *mask* (掩膜). Map extents are rectangular in shape; however, our study area may have an irregular shape (e. g. watershed, island, administrative boundary). A mask is a layer that contains the value No Data (or sometimes zeros) for cells outside of the study area. In this way the analysis is only conducted on cells within the study area.

8.3.2 Reclassification

Reclassifying converts cell values in the input layer to new values in the output layer. Thus it can be used for simplifying (or generalizing) a data layer (converting raw data into classes), and bringing values to a common scale-often converting to values in the range of 0−10 (where zero has no value and 10 has the highest value).

Raster data structure is a type of organization of GIS data storage, it is also called *grid data* (网格数据) or raster data. Pixel is the basic unit of rater data. Pixels are arranged by columns and rows with different gray values to form a gray level matrix. Raster data structure is a common spatial data format with the advantages of simple and direct structure, easy to operate and process.

Reclassification (重分类) is a process of taking input cell values and replacing them with new output cell values. Reclassification is often used to simplify or change the interpretation of raster data by changing a single value to a new value, or grouping ranges of values into single values — for example, assigning a value of "a" to cells that have values of 1 to 20, "b" to cells that range from 21 to 40, and so on.

Reclassification based on raster data is the most common method of data processing. The method is to reclassify original data and output new data. In accordance with different user requirements, types of reclassification include *replacing old data with new data* (新值替换), rearranging and *merging old data* (旧值合并), *reclassifying* (重分类) data with classification system, *setting specific data with null value* (空值设置).

Geographic objects are always changing. Therefore, old data often needs to be replaced for the sake of presenting realistic attributes. In the process of raster data operation, sometimes it is necessary to merge several classes to one class, such as to merge *paddy field* (水田) with *dry land* (旱地) together to form new class of *farmland* (耕地), to merge "river" with "lake" to form new class of "water". Sometimes, classified data needs to be reclassified with a new classification system. In particular cases, some data needs to be set with null value to compute. for example, data that isn't needed in the process of computing should be set as null value during analysis of mask.

The vector equivalent is to reclassify the attribute data in the table, and then dissolve common boundaries based on this generalization.

8.3.3 Raster Calculator

The raster calculator is a powerful and flexible analytical function. Sometimes, the ability to conduct mathematical operations on the values of raster cells is referred to as *map algebra* (地图代数). The functionality can be divided into two broad groups: Functions and Operators.

Mathematical functions are *arithmetic* (算数的), *logarithmic* (对数的), *trigonometric* (三角法的) and *power* (幂的) on a single raster layer. For instance, elevation is converted fromfeet to meters. Mathematical operators work on one or more layers. These operators evaluate the coincident cells (cells that overlap) from two or more layers and can be divided into three categories: arithmetic (e. g. +, -, *, /), Boolean (e. g. AND, OR, NOT, XOR) and relational (e. g. <, >, <=, >=, =, <>).

8.3.4 Interpolation (插值)

Many geographic phenomena vary continuously over the landscape, such as elevation, land cover, temperature, air pressure, soil depth, level of ground water, etc. This type of phenomena can be represented by point values (e.g. spot heights) or isolines (e.g. contours). However, the raster grid is an excellent data model for this type of data, as each cell is independent of the rest and can contain a unique value.

Source data often comes from point samples, whereby measurements are recorded at specific locations. Estimating the unsampled value at a location is a matter of interpolating a value based on the surrounding sample points. To determine the unsampled value at a point, greater weight should be given to the closer sample points (First Law of Geography).

8.3.5 Density Surface

Simply, GIS density is the quantity of phenomena distributed over per unit of area of a surface. For example, the population density means the number of people per square kilometer (person/km^2), Leaf Area Index (*LAI*, 叶面积指数) is defined as the one-sided green leaf area per unit ground surface area (m^2/m^2). In all similar cases, the end result is a raster data layer where the pixels contain density values (i.e. the number of something per square meter or square kilometre). The source data is usually in the form of points and the point values are distributed across the landscape.

There are simple and kernel density estimation methods. The simple method is a counting method, whereas the kernel method is based on a probability function and offers options in terms of how density estimation is made. To use the simple density estimation method, we can place a raster over a point distribution, tabulate points that fall within each cell, sum the point values, and estimate the cell's density by dividing the total point value by the cell size. A circle, rectangle, wedge, or ring based at the center of a cell may replace the cell in the calculation. Kernel density estimation associates each known point with a kernel function for the purpose of estimation. Expressed as a bivariate probability density function, a kernel function looks like a "bump", centering at a known point and tapering off to 0 over a defined bandwidth or window area. The kernel function and the bandwidth determine the shape of the bump, which in turn determines the amount of smoothing in estimation. Kernel density estimation uses the same input as for the simple estimation method. Density values in the raster are expected values rather than probabilities. Kernel density estimation usually produces a smoother surface than the simple estimation method does. Also, a larger bandwidth produces a smoother surface than a smaller bandwidth.

8.3.6 Surface

Surface analyses are fairly intuitive and include:

(1) Slope

This is a measure of change rate (e. g. steepness of slope). In fact, slope is the rate of maximum change in z-value from each cell of a raster surface. It can be upward or downward. Typically expressed as a percent, slope corresponds to vertical distance, divided by horizontal distance, multiplied by 100... Slope can also be expressed as an angle, which gives the amount of *deviation* (偏离) from the flat as a number of degrees. Conversions between slope percent and slope angle can be done using a scientific calculator and the *inverse tangent* (arctan, 反正切) function. Essentially, the slope angle is the inverse tangent of the slope percent (with slope percent expressed in decimal).

(2) Aspect

The aspect identifies the downslope direction of the maximum rate of change (steepest downslope direction) in value from each cell to its neighbors. Aspect can be seen as the slope direction (e. g. which way the hill is facing). The values of the output raster will be the *compass* (指南针) direction of the aspect and measured in degrees (azimuth 0–360).

(3) Contour

Creates a line feature class of contours (isolines) from a raster surface. Contours are lines that connect locations of equal value in a raster dataset that represents continuous phenomena such as elevation, temperature, *precipitation* (降水量), pollution, or *atmospheric pressure* (大气压力). The line features connect cells of a constant value in the input. Contour lines are often generally referred to as isolines but can also have specific terms depending on what is being measured. Some examples are *isobars* (等压线) for pressure, *isotherms* (等温线) for temperature, and *isohyets* (等降水量线) for precipitation.

(4) Viewshed (视域)

The output from this function shows the visible raster surface in the specified target locations (i. e. a given observation point or a set of observer features). The analysis uses the elevation value of each cell of the *DEM* (数字高程模型) to determine visibility to or from a particular cell.

(5) Hillshade (晕渲)

This function will create a *shaded relief* (地形阴影) from a surface raster by considering the *illumination source* (照明源) angles and shadows. This is done by placing a light source (the sun) in the sky. This placement of the light source includes direction and height-both given in degrees (i. e. horizontal and vertical angle). The output is quite often combined with the elevation layer to enhance the visualization of the surface. It can also be used in analysis to determine areas in light/shadow at any given time of the day (and any time of the year).

(6) Cut/Fill (挖/填)

Calculates the volume change between two surfaces. The cut-and-filloperations enable us to create a map based on two input surfaces—before and after—displaying the areas and volumes of surface materials modified by the removal or addition of surface material.

(7) Curvature (曲率)

Calculates the curvature of a raster surface, optionally including *profile and plane curvatures* (剖面曲率和平面曲率). The primary output is the curvature of the surface on a cell-by-cell basis, as fitted through that cell and its eight surrounding neighbors. Curvature is the *second derivative* (二次导数) of the surface, or the slope-of-the-slope. Two optional output curvature types are possible: the profile curvature is in the direction of the maximum slope, and the plane curvature is perpendicular to the direction of the maximum slope.

8.3.7 Distance Analyses

Distance in the vector world was a fairly straightforward concept. Examples include: measuring the distance between two (or more) points, determining the length along a road network, and creating buffers around features.

In raster we are still able to interactively measure the distance between cells. Distances may be expressed as physical distances or cost distances. The physical distance measures the Euclidean distance of cells, whereas the cost distance measures the cost for traversing the physical distance. Euclidean distance is calculated from the center of the source cell to the center of each of the surrounding cells. The Euclidean algorithm works as follows: for each cell, the distance to each source cell is determined by calculating the hypotenuse with x_max and y_max as the other two legs of the triangle. This calculation produces the true Euclidean distance, rather than the cell distance. The shortest distance to a source is determined, and if it is less than the specified maximum distance, the value is assigned to the cell location on the output raster. The cost distance calculates the least accumulative cost distance for each cell to the nearest source over a cost surface.

8.3.8 Map Algebra (地图代数)

Map algebra is a language specifically designed for geographic cell-based system and provides the basis for cartographic modeling. It means mathematical combinations of rasterlayers by several types of functions, which can be applied to one or multiple layers. Map algebra uses an expression to link the input and the output. Besides the input and output, the expression can be composed of GIS tools, mathematical operators, and constants. GIS tools can include not only the basic tools of local, focal, zonal, and distance measure operations but also special tools such as Slope for deriving a slope raster from an elevation raster. The expression, Slope (emidalat, degree), can be used to derive "slope_d", a slope raster measured in degree, from "emidalat" an elevation raster.

(1) Local Functions (point functions)

Local functions apply their calculations to a single cell location before calculating the next location, until all cells have been processed (Figure 8-6). In other words, local functions work on every single cell in a raster layer and cells are processed without reference to surrounding cells. To perform the calculation, the local function only needs to know the values at the location for a sin-

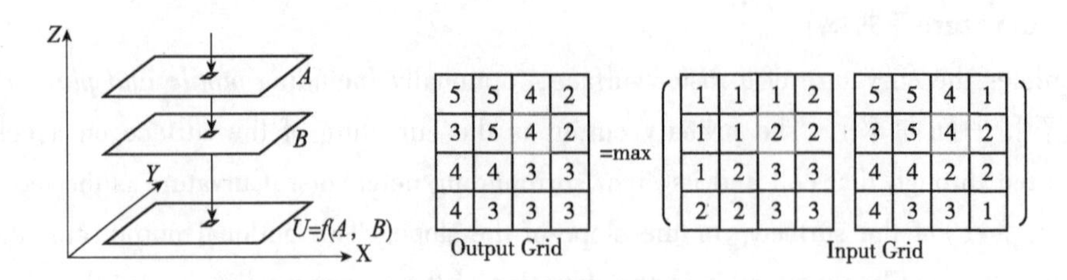

Figure 8-6　Spatial Context and Example of Local Functions

gle raster or for multiple rasters, as well as, in some cases, a comparison value. A local function results in a new grid as a function of one or more input grids. Operations can be arithmetic, trigonometric, *exponential* (指数的), logical or logarithmic functions. The generic form of local functions is as: $U = f(X_1, X_2, \ldots)$. For example: $U = \sin(X_1)$, $U = \min(X_1, X_2, \ldots)$, $U = \text{Merge}(X_1, X_2, \ldots)$, $U = \text{Select value} >= 10\ (X_1)$, and so on.

(2) Focal Functions (neighborhood functions)

Focal functions are also referred to as neighborhood functions because the value calculated for each cell in the result is a function of the values of all the cells on a data raster within a *predefined* (预定义) neighborhood around that cell. Neighborhoods can be defined by rectangles, circles, *wedges* (楔形), *doughnut shapes* (*annulus*, 环状), etc. This function explicitly makes use of some kind of spatial associations in order to determine the value for the locations on the new output grid. Every focal operation requires at least three basic parameters: ①target location(s) (neighborhood focus); ②a specific neighborhood around each target; ③a function to be performed on the elements within the neighborhood.

A neighborhood configuration determines that cells surrounding the processing cell should be used in the calculation of each output value. To complete a neighborhood function on a raster, the focal cell is moved from one cell to another until all cells are visited. Neighborhood functions can return the maximum, minimum, mean, *standard deviation* (标准偏差), sum, median, and range of values within the immediate or extended neighborhood, as well as tabulation of measures such as majority, minority, and variety. These statistics and measures are the same as those from local functions with multiple rasters. Figure 8-7 shows the results of the neighborhood processing for calculating a mean statistic within a predefined circle neighborhood. A block function is a neighborhood function that uses a rectangle (block) and assigns the calculated value to all block

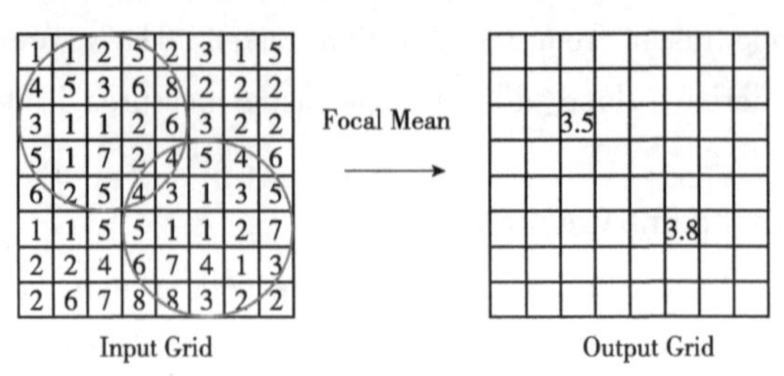

Figure 8-7　Example of Focal Functions

cells in the output raster. Therefore, a block function differs from a regular neighborhood function because it does not move from cell to cell but from block to block.

An important application of neighborhood function is data simplification. The moving average method, for instance, reduces the level of cell value fluctuation in the input raster. The method typically uses a 3-by-3 or a 5-by-5 rectangle as the neighborhood. As the neighborhood is moved from one focal cell to another, the average of cell values within the neighborhood is computed and assigned to the focal cell. The output raster of moving averages represents a generalization of the original cell values. Neighborhood functions can also be important to studies that need to select cells by their neighborhood characteristics.

Neighborhood functions are common in image processing. These operations are variously called filtering, convolution, or moving window operations for spatial feature manipulation.

(3) Zonal Functions (regional functions)

A zone is an area or a region distinguished from adjacent parts by a distinctive feature or characteristic. Zones define cells that share a common characteristic, and cells in the same zone don't have to be contiguous. A contiguous zone includes cells that are spatially connected, whereas a noncontiguous zone includes separate regions of cells. A watershed raster is an example of a contiguous zone, in which cells in the same watershed are spatially connected. A land use raster is an example of a noncontiguous zone, in which one type of land use may appear in different parts of the raster.

The general rules for setting up zones are: ①it is easier to aggregate than disaggregate; ②too many zones increase processing difficulties, and may produce insufficient data *degrees of freedom* (自由度); ③no matter how careful you select zones, they will be unsatisfactory to some extent because real boundaries are *fuzzy* (模糊) and overlap; ④recall that the reason for disaggregation and *stratification* (分层) is to partition variance.

Zonal functions operate on properties of the region (or zone) to which a given cell belongs [区域函数作用于某一指定象元所属区域(或地带)的属性]. Zonal functions produce a new grid of zones that summarize the values of a data grid over the areas covered by the zones in a zone grid (Figure 8-8). Zonal functions are similar to focal functions except that the definition of the neighborhood in a zonal function is the configuration of the zones or features of the input zone dataset, not a specified neighborhood shape. However, zones do not necessarily have any order or specific shapes. Each zone can be unique. Zonal functions return the mean, sum, minimum,

Input Grid

49	46	41	25	18	19
88	44	12	9	56	66
33	29	98	42	48	42
26	37	46	38	71	15
12	65	22	16	62	31
90	53	17	17	51	11

Zone Grid

a	a	a	b	b	b
a	a	a	b	b	b
d	a	b	b	b	c
d	d	b	b	c	c
d	d	e	e	c	c
d	d	e	e	e	e

Output (Zonal Sum) Grid

309	309	309	465	465	465
309	309	309	465	465	465
316	309	465	465	465	224
316	316	465	465	224	224
316	316	134	134	224	224
316	316	134	134	134	134

Figure 8-8 Example of Zonal functions

maximum, or range of values from the first dataset that fall within a specified zone of the second.

Characteristics of zonal functions: ①do not change boundaries of regions; ②change attribute values for each region (or zone) according to its statistics or user's specification; ③useful for understanding spatial distribution of objects, quantitative measurement of shapes, statistical properties of objects and spatial associations.

Zonal functions with two rasters can generate useful descriptive statistics for comparison purposes. For example, to compare topographic characteristics of different soil textures, we can use a soil raster that contains the categories of sand, loam, and clay as the zonal raster and slope, aspect, and elevation as the input rasters.

(4) Global Functions

Global functions perform operations based on the input of the entire grid, such as distance measures, flow directions (See Chapter 9) or flow accumulation. It is useful when we want to work out how cells "relate" to each other.

8.3.9 Other Raster Data Operations

Local, neighborhood, zonal, and distance measure operations cover the majority of raster data operations. Some operations, however, do not fit well into the classification. There are some other raster data operations.

①Mosaic: Mosaic is a tool that can combine multiple input rasters into a single one. If the input rasters overlap, a GIS package typically provides options for filling in the cell values in the overlapping areas.

②Extraction: Raster data extraction creates a new raster by extracting data from an existing raster. The operation is similar to raster data query. A graphic object for raster data extraction can be a rectangle, a set of points, a circle, or a polygon. The extract-by-attribute operation creates a new raster that has cell values meeting the query expression.

③Generalization: Several operations can generalize or simplify raster data. One operation is resampling, which can build different pyramid levels (different resolutions) for a large raster data set. Aggregating is similar to resampling in that it creates an output raster with a larger cell size. Some data generalization operations are based on zones, or groups of cells of the same value.

8.4 Comparison between Vector-based Data Analysis and Raster-based Data Analysis

Vector data analysis and raster data analysis represent the two basic types of GIS analyses. Each GIS project is different in terms of data sources and objectives. Moreover, vector data and raster data can be easily converted each other. We must therefore choose the efficient and appropriate type of data analysis. As the two most common operations in GIS, overlay and buffering are used to compare vector-based and raster-based operations.

A local operation with multiple rasters is often compared to a vector-based overlay operation. The two operations are similar in that they both use multiple data sets as inputs. But important differences exist between them. First, to combine the geometries and attributes from the input layers, a vector-based overlay operation must compute intersections between features and insert points at the intersections. This type of computation is not necessary for a raster-based local operation because the input rasters have the same cell size and area extent. Even if the input rasters have to be first resampled to the same cell size, the computation is still less complicated than calculating line intersections. Second, a raster-based local operation has access to various tools and operators to create the output whereas a vector-based overlay operation only combines attributes from the input layers. Any computations with the attributes must follow the overlay operation. Given the two reasons above, raster-based overlay is often preferred for projects that involve a large number of layers and a considerable amount of computation. Although a raster-based local operation is computationally more efficient than a vector-based overlay operation, the latter has its advantages as well. A vector-based overlay operation can combine multiple attributes from each input layer. Once combined into a layer, all attributes can be queried and analyzed individually or in combination. A vector-based overlay operation is therefore more efficient than a raster-based local operation if the data sets to be analyzed have a large number of attributes that share the same geometry.

A vector-based buffering operation and a raster-based physical distance measure operation are similar in that they both measure distances from selected features. But they differ in at least two aspects. First, a buffering operation uses x and y coordinates in measuring distances, whereas a raster-based operation uses cells in measuring physical distances. A buffering operation can therefore create more accurate buffer zones than a raster-based operation can. This difference on accuracy can be important, for instance, in implementing riparian zone management programs. Second, a buffering operation is more flexible and offers more options. For example, a buffering operation can create multiple rings (buffer zones), whereas a raster-based operation creates continuous distance measures. Additional data processing (e.g. reclassification or slice) is required to define buffer zones from continuous distance measures. A buffering operation has the option of creating separate buffer zones for each selected feature or a dissolved buffer zone for all selected features. It would be difficult to create and manipulate separate distance measures using a raster-based operation.

8.5 Geographical Analysis Procedure

General procedure of geographical analysis is:

①Establish the objectives and criteria for the analysis (确定分析目标和规则). Good decisions are based on clear objectives. They should be specific, measurable, agreed, realistic and time-dependent. It is critically important to decide on how to compare different options' contribution to meeting the objectives. This requires the selection of criteria to reflect performance in meeting the objectives. Each criterion must be measurable, in the sense that it must be possible to assess, at least in a qualitative sense, how well a particular option is expected to perform in relation to the criterion. Therefore, in this step, we need to define the problem and then identify a sequence of operations to produce meaningful results.

②Prepare the data for spatial operations (准备用于空间操作的数据). The step is tocollect and acquire geographic information that will help to answer questions mentioned in the first step. We need to prepare all map coverages for the proposed data analysis, add one or more attributes to these coverages in the database if necessary.

③Perform the spatial operations (执行空间操作). Perform the spatial operations and combine the coverages, e. g. creating buffering zones around features, manipulating spatial features and performing polygon overlay.

④Prepare the derived data for tabular analysis (准备用于表格分析的派生数据). Make sure the feature attribute table contains all the items needed to hold the new values to be created.

⑤Perform the tabular analysis (进行表格分析). Calculate and query the relational database using the model defined in step 1.

⑥Evaluate and interpret the results (评估并解释结果). We need toformulate reasonable answers to the geographic questions based on our analysis and interpretation of the information gathered, and examine the results and determine whether the answers are valid. Simple map displays and reports can help in this evaluation.

⑦Refine the analysis and repeat the analysis if necessary (精炼分析结果并且必要时重新进行分析).

Vocabulary

advent[ˈædvənt] *n.* 出现
configuration[kənˌfɪɡəˈreɪʃn] *n.* 结构；布局；形态；格式塔心理完形
corridors[ˈkɒrɪdɔːz] 走廊
crux[krʌks] *n.* 关键
deceive[dɪˈsiːv] *v.* 欺骗
donuts *n.* 甜甜圈(donut 的复数)；油炸圈饼
explicit[ɪkˈsplɪsɪt] *adj.* 明确的
feature[ˈfiːtʃə] *n.* 特征
focal[ˈfəʊk(ə)l] *adj.* 焦点的
generalisation[ˌdʒenərəlaiˈzeiʃən] *n.* 一般化；归纳
illuminate[ɪˈlumɪnet] *vt.* 阐明
implicit[ɪmˈplɪsɪt] *adj.* 暗示的
interpret[ɪnˈtɜːprɪt] *vi.* 解释 *vt.* 说明
intuitively[inˈtju：itivli] *adv.* 直观地
invariant[ɪnˈveərɪənt] *adj.* 不变的

isoline[ˈaɪsəʊlaɪn] *n.* 等直线(等斜褶皱)
maintenance[ˈmeɪnt(ə)nəns] *n.* 维护
manipulation[mə, nɪpjʊˈleɪʃ(ə)n] *n.* 操作
mathematical[mæθ(ə)ˈmætɪk(ə)l] *adj.* 数学的
matrix[ˈmeɪtrɪks] *n.* 矩阵
merge[mɜːdʒ] *vt.* 合并
pattern[ˈpætən] *n.* 模式
planar[ˈpleɪnə] *adj.* 平面的；二维的
pursuit[pəˈsjuːt] *n.* 追求
quantitative[ˈkwɔntɪ, tətɪv] *adj.* 定量的
query[ˈkwɪərɪ] *n.* 疑问；质问
raster[ˈræstə] *n.* 光栅
registration[redʒɪˈstreɪʃən] *n.* 注册
reservoir[ˈrezəvwɑː(r)] *n.* 水库
so-called 所谓的
spectral[ˈspektr(ə)l] *adj.* 光谱的
splice[splaɪs] *vt.* 拼接
straddle[ˈstrædl] *v.* 跨坐；把两腿叉开；观望 *n.* 跨坐；观望；指定价格的交易
successor[səkˈsesə] *n.* 继承者
superimpose[ˌsuːp(ə)rɪmˈpəʊz] *vt.* 重叠；附加
synthesis[ˈsɪnθɪsɪs] *n.* 综合；综合体
tabular[ˈtæbjʊlə] *adj.* 列成表格的
topological[tɔpəˈlɔdʒikəl] *adj.* 拓扑的

Questions for Further Study

1. Definitions of spatial analysis and topological overlay.
2. What is the role of vector data-based overlay?
3. What is the difference between vector data-based topological overlay and raster data-based topological overlay?
4. Illustrate the use of buffering analysis.
5. Analyze and explain the methods and steps of how to select a suitable site for industrial factory based on raster data (Supposing that only factors of land use, slope, population density and nature reserve are in consideration).

Chapter 9

Digital Terrain Model and Digital Elevation Model

A *Digital Terrain Model* (DTM) is an ordered *array* (阵列) of numbers that represents the spatial distribution of terrain characteristics. In the most usual case, the spatial distribution is represented by the horizontal coordinate value *X*, *Y* and the value *Z* recording the terrain characteristic or elevation. DTMs are the most important spatial information in a geodatabase and a major constituent of geographical information processing and analyzing. They provide not only a basis for numerous applications in the terrain-related fields such as surveying and mapping, resources and environment, disaster prevention and control, national defense, but also a method to model, analyze and display phenomena related to topography (数字地形模型是表示地形空间分布特征的有序数字阵列。通常情况下，人们使用一对 *X*，*Y* 水平坐标值和记录地表特征或高程的 *Z* 值表示地形的空间分布状态。数字地形模型是地理数据库中最重要的空间信息资料，也是地理信息处理和分析的重要成分。它不仅在与地形相关的诸如测绘、资源与环境、灾害防治、国防等领域中广泛应用，并且为建模、分析和显示与地形有关的各种现象提供了一种方法).

9.1 DTM and DEM

The term Digital Terrain Model has its origin in work performed by Prof. Charles L. Miller at *Massachusetts Institute of Technology* (麻省理工学院) about 1955-1960, and the objective was to *expedite* (加快，促进) highway design by digital computation based upon photogrammetrically acquired terrain data. A DTM may be understood as a mathematical representation (model) of the ground surface, most often in the form of a regular grid, in which a unique elevation value is assigned to each pixel (i. e. 2D point). And it can be viewed as a topographic model of the *bare earth* (裸露地表) that enables users to *infer* (推断) terrain characteristics. A DTM has had vegetation, buildings, and other cultural features digitally removed, leaving just the underlying terrain. It is a 2. 5D rather than a true 3D model of the terrain (Weibel et al. , 1991) because DTM fails to describe vertical terrain features (e. g. cliffs) with a single height value assigned to each 2D point [通常，由于 DTM 中每个二维点坐标仅被赋予一个单一高程，无法表示垂直地形特征(如悬崖)，因此，DTM 不是地表形态的真三维模型而是 2.5 维的].

In practice, different definitions of DTM are put forward by groups in different countries, such as DGM (Digital Ground Model), DHM (Digital Height Model), DEM (Digital Elevation Model), and DTEM (Digital Terrain Elevation Model). Generally, these definitions are often assumed to be *synonymous* (同义的) although sometimes they are distinctly different. Li (1990) has made a comparative analysis of these differences as follows:

①Ground: The solid surface of the earth; a solid base or foundation; a surface of the earth; bottom of the sea; etc.

②Height: Measurement from base to top; elevation above the ground or recognized level, especially that of the sea; distance upwards; etc.

③Elevation: Height above a given level, especially that of the sea; height above the horizon; etc.

④Terrain: A tract of land with regards to its natural features; an extent of ground, region, territory; etc.

Commonly, a DTM is simply an ordered set of sampled data points with known x, y, z coordinates that represent the spatial distribution of various types of information on the terrain. The mathematical expression of DTM is as:

$$K_p = f_k(u_p, v_p) \quad k = 1, 2, 3, \cdots, m; \; p = 1, 2, 3, \cdots, n. \tag{9-1}$$

where, K_p is the attribute value of the kth terrain information at the location of point p, which can be a single point, but is usually a small area centered by p; (u_p, v_p) is the 2D coordinate pair of point p; m is the total number of terrain information types ($m \geqslant 1$); n is the total number of sampled points. DTM is a more generic term for any digital representation of a topographic surface (对于地形表面的任何一种数字表达方式而言，DTM 是一个更通用的术语). Accordingly, a DTM may be used as a digital model of any single-valued surface, e. g. geological horizon, air temperature or air pressure, population density, and so on. Especially, when m equals 1 and the terrain information is "height", the result is the mathematical expression of DEM. Therefore, DEM is just a subset of DTM. DEM has *superseded* (取代) other terms such as DGM, DHM, and DTEM to refer to terrain models with elevation information only. Most of the data providers (USGS, ERSDAC, CGIAR, Spot Image) use the term DEM as *a generic term* (通用术语) for DTM. All datasets captured with satellites, airplanes or other flying platforms are originally DTMs, such as SRTM (Shuttle Radar Topography Mission, 航天飞机雷达地形测绘任务) or the ASTER GDEM (Advanced Spaceborne Thermal Emission and Reflection Radiometer Global Digital Elevation Model, 先进星载热发射和反射辐射成像仪全球数字高程模型).

The term DEM can refer to one of the following: ①A digital representation of a continuous variable over a two-dimensional surface by a regular array of z values referenced to a common datum. DEMs are typically used to represent terrain relief and frequently comprise a foundational layer in any *archaeological*（考古学的）GIS database. ②An elevation database for elevation data by map sheet from the National Mapping Division of the U. S. Geological Survey (USGS). ③The format of the USGS digital elevation data sets (ESRI, 1996).

The process for the construction of a DTM surface is called digital terrain modeling. It is also a process of mathematical modeling. According to *Weibel* et al. (1990), digital terrain modeling encompasses the following general tasks (Figure 9-1):

Generation: Data collection and model construction.

Manipulation: Modification and *refinement*（细化）of the model.

Interpretation: Analysis and information extraction.

Visualization: Graphical *rendering*（渲染）of the terrain model.

Application: Development of appropriate application models for specific disciplines.

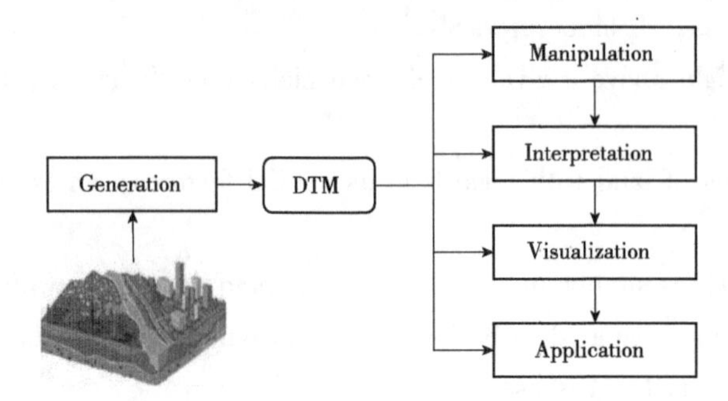

Figure 9-1　General Tasks of Digital Terrain Modeling

9.2　DTM Generation

9.2.1　Data Sources and Data Capture

The choice of data sources and terrain data sampling techniques is critical for the quality of the resulting DTM. Data for a DTM should consist of *observations*（观察值）about terrain elevations and additional information about phenomena that significantly influence the shape of the terrain surface (i. e. structural features such as drainage channels, ridges, and other surface discontinuities). Most DTM data are derived from three alternative sources: ground surveys, existing topographic maps, or remote sensing（大部分数字地形模型数据源于地面测量、现有的地形图或遥感）. Other sampling methods *occasionally*（偶尔地）used include *radar or laser altimetry*（雷达或激光测高计）, and *sonar*（*for subaquatic terrain*）[声波定位仪(用于水下地形)]. Data for geological models are obtained from either *borehole records*（钻孔记录）or *seismic surveys*（地震探测）.

The horizontal and vertical location of points on the earth's surface with an accuracy of a few millimeters can be geolocated with traditional and high-tech ground survey techniques, such as *theodolites*（*instruments for measuring angles in horizontal and vertical planes*, 经纬仪）, notebooks, and triangulation methods (calculating distances and angles between points), electronic theodolites, *total stations* (全站仪), and *electronic distance measuring* (EDM, 电子测距仪). The complexity and cost of surveying with such equipment require dedicated surveying teams, which often go beyond the means of small mapping projects.

In situations where individuals, agencies have no access to DTM data with expensive (and more accurate) methods, manual or semi-automated digitizing of topographic maps and cartographic maps can convert paper or raster images of these maps into vectors and be used as inputs to *interpolation* (内插) algorithms for creating a DTM surface, although these methods represent a worst-case scenario and are no longer the main source of DTM data.

In contrast to the methods shown above, remote sensing, i. e. the interpretation of image data acquired from airborne or satellite platforms, can rapidly cover large areas and will lead to further cost reduction and time reduction in DTM generation. The platforms for remote sensing mainly include: ①photogrammetric/stereo methods (encompassing both airborne and satellite); ②*laser* (激光, mostly airborne at present, but will be from satellites in the future); and ③radar (both airborne and satellite, using *interferometry*, 干涉法).

9.2.2 DTM Construction

The process of terrain data capture generates a set of relatively unordered data elements. To construct a comprehensive DTM, it is necessary to establish the topological relations between the data elements, as well as an interpolation model to approximate the surface behavior.

(1) Data Distribution Characteristics and Data Structures

Under the characteristics of spatial distribution, the data of DTM can be divided into two categories: *grid data* (格网数据) and *discrete data* (离散数据). Suppose the earth's surface can be divided into an ordered set of regular grids and each grid is uniform in size and shape. The row number and column number of the grid matrix indicate the (x, y) coordinates, while the z value can be any one of the 3D terrain information such as elevation, gradient, aspect, etc. The size of the grid represents the data accuracy. However, not all the observations can be collected according to the regular grid due to limitations in observation instruments or the landform. Therefore, discrete data is needed and the (x, y) coordinates are decided by geospatial location.

The original data must be structured to enable handling bysubsequent terrain modeling operations. The overwhelming majority of DTMs conform to one of the two data structures: Rectangular Grid or TIN (Triangulated Irregular Network) [大多数 DTM 遵从两种数据结构: 矩形格网或 TIN(不规则三角网)] (Figure 9-2).

Rectangular grids present a matrix structure that records topological relations between data points implicitly. Since this data structure reflects the storage structure of digital computers, the handling of elevation matrices is simple and grid-based terrain modeling algorithms thus tend to be relatively straightforward. However, the point density of regular grids cannot be adapted to the complexity of the relief. Thus, an excessive number of data points are used to represent the

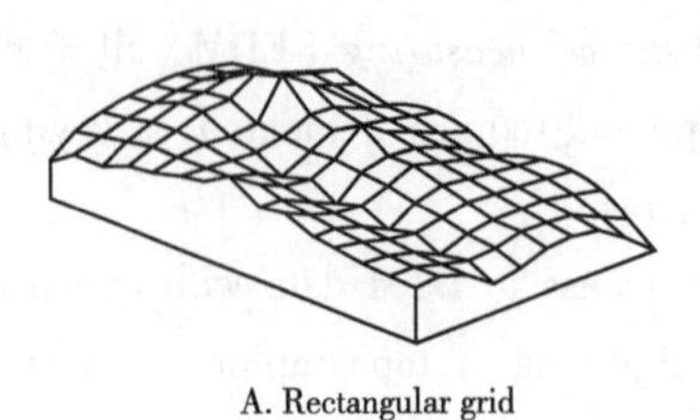

A. Rectangular grid

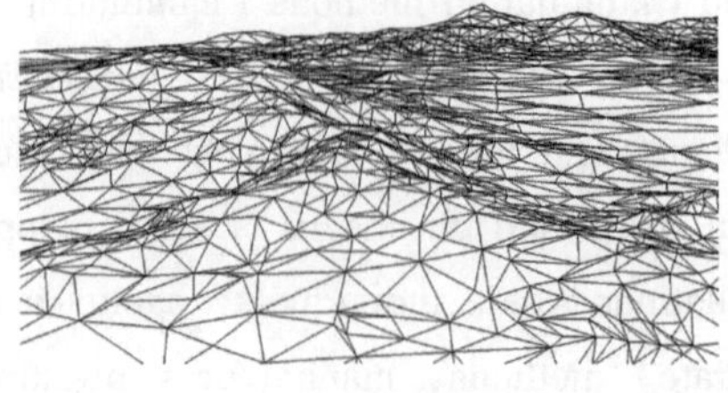

B. TIN

Figure 9-2 DTM Data Structure

terrain in a required level of accuracy. Also, rectangular grids cannot describe the structural features in comparison with the topographic features; extensions to the basic model have to be added for this purpose.

The TIN structure is based on triangular elements, with vertices at the sample points. Structural features can easily be incorporated into the data structure. Consequently, TINs can reflect adequately the variable density of data points and the *roughness* (粗糙度) of terrain. Fewer points are needed for a DTM of certain accuracy. However, topological relations have to be computed or recorded explicitly. Thus, TINs become more complex and also more difficult to handle. Not all grid-based algorithms have an efficient counterpart in TIN structures.

The TIN model represents a terrain as a set of contiguous, *non-overlapping* (不重叠) triangles. Within each triangle, the terrain is represented by a plane. The triangles are made from a set of points (mass points). Mass points can occur at any location. The more carefully selected, the more accurate the model of the terrain. The ground returns well-placed mass points occur where there is a major change in the shape of the earth surface, for instance, at the *peak* (峰值) of a mountain, the floor of a *valley* (山谷), or at the edge (top and bottom) of cliffs.

The TIN model is attractive because of its simplicity and economy and is a significant alternative to the regular grid of the GRID model (Table 9-1).

Table 9-1 Comparison Between Two Main Data Structures of DTM

	TIN	Grid
Advantages	ability to describe the terrain at different levels of resolution efficiency in storing data	easy to store and manipulate easy integration with raster databases the smoother, more natural appearance of derived terrain features
Disadvantages	require visual inspection and manual control of the network in many cases	inability to use various grid sizes to reflect areas of different complexity of relief

(2) Interpolation

In digital terrain modeling, interpolation serves the purpose of estimating elevations (or other terrain information) in regions where no data exists. Interpolation is mainly used for computing.

①Elevations (z) at single point locations.

②Elevations (z) of a rectangular grid from original sampling points.

③Locations (x, y) of points along contours (in contour interpolation).

④Resampling.

A popular way to classify interpolation models is by the influence range of the data points involved. Global methods in which all sample points are used for interpolation may be distinguished from local methods (piecewise methods, 时域分段法) in which only data points nearby are considered. Because topographic surfaces are non-stationary and non-periodic, the use of overly distant points may *deform* (使变形) the interpolated surface. For DTM with sample points of sufficient quality and density, a local and exact interpolation method on surface *patches* (补丁) is widely considered satisfactory.

Characteristics and *peculiarities* (特性) of DTM interpolation from topographic samples:

①No "best" interpolation algorithm is superior to all others and appropriate for all applications.

②The quality of the resulting DTM is determined by the distribution and accuracy of the original data points (i. e. the sampling process), and the adequacy of the underlying interpolation model (a hypothesis about the behavior of the terrain surface).

③The most importantcriteria for selecting a DTM interpolation method are the degree to which ①structural features can be taken into account, and ②the interpolation function can be adapted to the varying terrain character.

④Suitable interpolation algorithms must adapt to the character of data elements (type, accuracy, importance, etc.) as well as the context (i. e. the distribution of data elements). Satisfactory solutions exist for the interpolation of relatively well-selected and dense topographic samples of estimating elevations in regions where no data exist.

(3) Triangulation

Some of the more widely used local interpolation procedures are based on triangulation: interpolation is achieved by locally fitting *polynomials* (多项式) to triangles. Furthermore, triangulation is used to construct TIN DTM. Triangulation thus serves two purposes in terrain modeling: the basis for TIN data structures and the basis for interpolation (三角测量法在地形建模中起着两种作用: TIN 数据结构的基础和数据插值的基础). Commonly, the two-or three-step procedure in the TIN DTM construction is probably: ①a TIN is constructed based on mass points; ②it is then used for interpolation so that a triangulation of irregular distributed points are reconstructed; ③it is then added elevation values to the point features (Figure 9-3).

If a triangulation of a set of points meets: ①the *circumcircle* (外接圆) of any of its triangles does not contain any other point in its interior; ②the triangles are as *equiangular* (等角) as possible, i. e. triangulations maximize the minimum angle of all the angles of the triangles in the triangulation, thus reducing potential numerical precision problems created by long skinny triangles; ③any point on the surface is as close as possible to a node, i. e. the sum of side lengths of the tri-

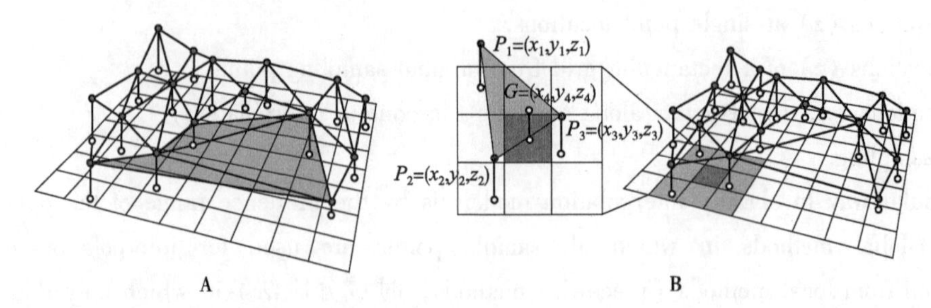

A. TIN model (red circles are pointing to an observed point cloud, red linesare TIN, and green triangles represent the removed TIN elements with long edges and consisting of a large area around the point cloud); B. TIN–based interpolation for DEM generation: blue lines are DEM grids, and the 3D coordinates of a white circle (G) are *x*/*y* location (*xg*, *yg*) and interpolated height (*zg*) from the vertices of the TIN element (P_1, P_2, and P_3).

Figure 9-3 Two-step TIN DTM Construction Procedure

(Cho, 2015)

angle tends to be the smallest one. Then it is said to be Delaunay triangulation. The Delaunay triangulation is bounded by the *convex hull* (凸包) and the triangulation is independent of the order the points are processed

The Delaunay triangulation with all the circumcircles and their centers (Figure 9-4D). Connecting the centers of the circumcircles produces theVoronoi diagram or *Thiessen polygon* (泰森多边形) (Figure 9-4E).

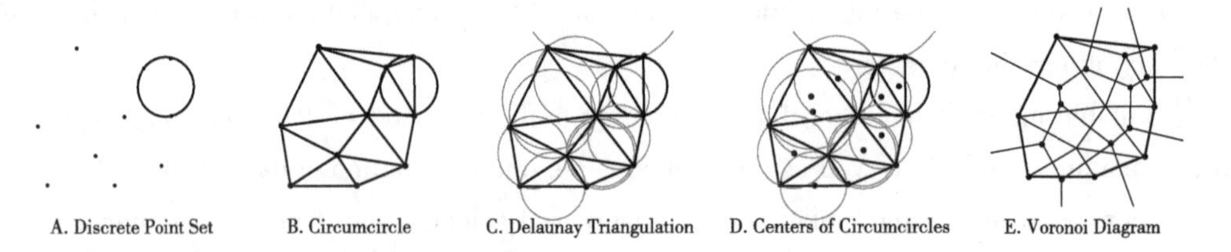

Figure 9-4 The Delaunay Triangulation of a Discrete Point Set in General Position Corresponds to the *Dual Graph* (对偶图) of the Voronoi diagram

9.3 DTM Manipulation

Along with DTM generation procedures, the DTM manipulation processes are of fundamental importance for the performance and flexibility of a DTM system. They are needed for the modification and refinement of existing models. DTM manipulation consists of DTM editing, filtering and merging, and the conversion between different data structures.

9.3.1 DTM Editing

DTM editing involves updating and error correction. An editor is required for interactive, selective modification of the properties of individual elements of a DTM. Edit operations for DTM elements should include: query, deleting, adding, moving, changing height, changing attribute, etc. For grid DTM, however, editing is essentially restricted to modifying elevations at grid points. If TIN is edited, algorithms are required for local adjustment of the network topology after points have been inserted or deleted.

9.3.2 DTM Filtering

DTM *filtering* (筛选) may serve two purposes: smoothing or enhancement of DTM, as well as data reduction.

Smoothing and enhancing filters for DTM is equivalent to *lowpass* (低通的) and *highpass filters* (高通滤波器). They are best applied to grid terrain models. Smoothing (i. e. applying lowpass filters) is to remove details, and make the DTM surface smoother. Enhancing (i. e. highpass filtering) has the opposite effect: surface discontinuities are emphasized, while smooth shapes are suppressed. Smoothing filters have been used to eliminate *blunders* (大错) (e. g. in photogrammetric data) and they may also be used to some extent for *cartographic generalization* (制图综合) of DTM. Enhancing (or highpass) filters have rarely been applied to DTM.

DTM filtering procedures are also used to reduce the data volume of DTM. A data reduction process of this kind may be desirable to eliminate redundant data points (e. g. within the digitizing tolerance), to save storage space and processing time, or to reduce the resolution of DTM (e. g. for comparison with other models). It may also be used as a pre-processing step in cartographic generalization of DTM, or to convert a grid DTM into a TIN.

9.3.3 DTM Joining and Merging

DTM may be combined either by joining adjacent models or by merging overlapping models.

For grid DTM, joining is only straightforward if the grids correspond to grid resolution, orientation, etc. Otherwise, aresampling process (for coarsening, densification, and/or reorientation) has to be used to establish continuity. The joining of TIN requires algorithms for connection and readjustment of the TIN along their borders (zipping) to patch the models together.

Merging DTMs poses problems at two levels. The first task consists of inserting all elements of one model into the other model (e. g. by incremental triangulation versus full re-triangulation). The second problem domain involves combining data sets with conflicting attributes (elevation and gradients) and varying degrees of accuracy. It is complex and represents a fundamental methodical problem.

9.3.4 Data Structure Conversion

The task of converting a DTM of a certain representation (e. g. TIN) into another structure (e. g. grid) can be handled by a combination of DTM generation and manipulation procedures.

Grid-to-TIN conversion may be handled by the same procedure used for data reduction of TIN. To be able to exploit the benefits of a TIN data structure (i. e. adapt the point density to the terrain complexity to save storage space), insignificant points must be discarded in the conversion process. There are also strong indications that this algorithm is more suitable for handling grid-to-TIN conversion than other approaches.

Other conversions are essentially equivalent to interpolation processes discussed in the preceding section. TIN-to-grid conversion is equivalent to TIN-based interpolation of a grid DTM. Contour-to-grid and contour-to-TIN conversion are special *variants* (变体) of interpolation tasks. Grid-to-contour and TIN-to-contour processes are solved by contour interpolation.

9.4 DTM Interpretation

Within a GIS, DTM is the most valuable as a basis for the extraction of terrain-related attributes and features. Information may be extracted in two ways: visual analysis of graphic representations (i. e. through visualization) and quantitative analysis of digital terrain data (i. e. through interpretation).Interpretation procedures, along with visualization functions, represent an important objective of GIS-related terrain modeling. The results of interpretation can be used as input to environmental impact studies, soil erosion potential models, hydrological runoff simulations, and many more applications.

The prime objective of DTM interpretation is the derivation of *geomorphometric* (地形的) parameters. Geomorphometric analysis may take two forms: general geomorphometry and specific geomorphometry.

9.4.1 General Geomorphometry

The most frequent use of general geomorphometry is the derivation of slope values from DTMs. The slope isclassified as a vector and defined by a *plane tangent* (相切平面) to the DTM surface at any given point, and comprises a quantity (gradient, 坡度), and a direction (aspect, 坡向). Slope gradient is defined as the maximum rate of change in altitude ($\tan Q$), and aspect (y) as the compass direction of this maximum rate of change (Figure 9-5). Apart from being displayed as slope maps, gradient and aspect are often used as numerical input to GIS models. Besides gradient and aspect (i. e. the first derivatives of the altitude surface), the second derivative (or rate of change of slope) convexity (or curvature, 曲率) is often used for geomorphological analysis. Convexity also has two components: *profile convexity* (剖面曲率), i. e. the rate of change of gradient; and *plan convexity* (平面曲率), the convexity of contours.

Slope gradient is a vector, i. e. it has a direction and length. A slope can be measured by its horizontal angle (slope) and azimuth (aspect). A common algorithm to calculate slope is the one generating slope and aspect values from its 'normal vector'. A normal vector of a surface is the vector perpendicular to the surface with the direction pointing away from the surface.

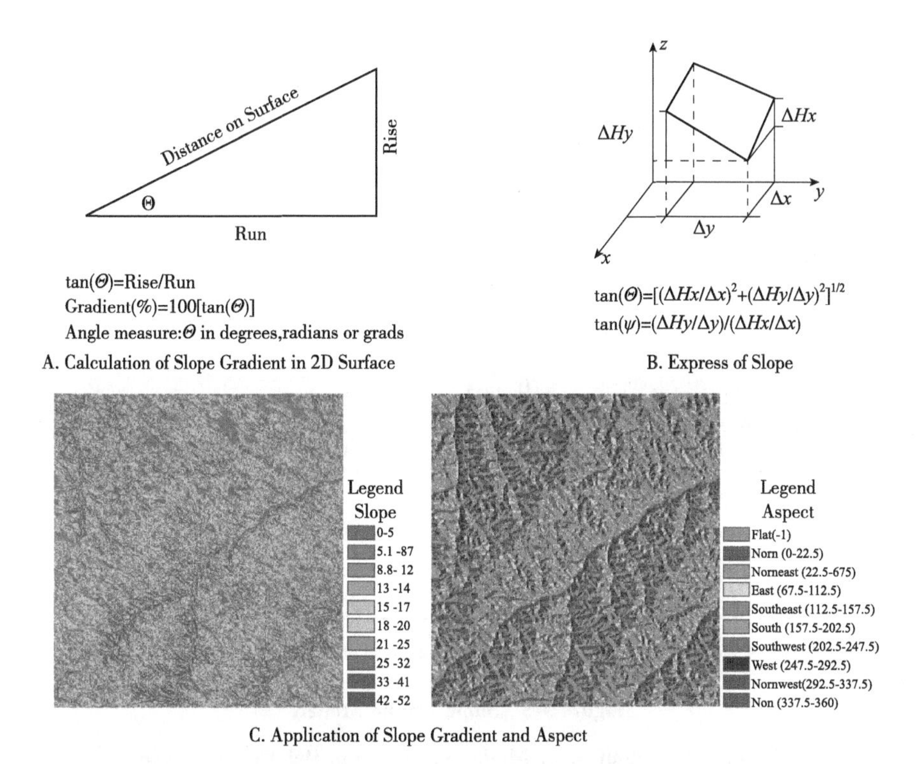

A. Calculation of Slope Gradient in 2D Surface

B. Express of Slope

C. Application of Slope Gradient and Aspect

Figure 9-5 Slope and Aspect Sketch

(Note: ΔHx and ΔHy—the local relief within the neighborhood along the west-east and south-north directions respectively; ΔHx andΔHy—the distance of the neighborhood along west-east and south-north directions)

Terminology used to describe slop angle is shown in Table 9-2.

Table 9-2 Terminology Used to Describe Slope Angle

Angle	Description	Angle	Description
0°	Plain	15°–25°	Steep
0°–30′	Slightly Sloping	25°–35°	Very Steep
2°–5°	Gently inclined	35°–55°	Precipitous
5°–15°	Strongly inclined	55° and greater	Vertical

Aspect refers to the direction of the normal vector in a clockwise direction from the north. It is usually measured in degrees clockwise from North, where 0 degree is due north, 90 degrees is due east, 180 degrees is due south, and 270 degrees is due west. Categories of aspect are listed as north (337.5°–22.5°), northeast (22.5°–67.5°), east (67.5°–112.5°), southeast (112.5°–157.5°), south (157.5°–202.5°), southwest (202.5°–247.5°), west (247.5°–292.5°) and northwest (292.5°–337.5°).

Slope profile (坡形) is defined by the shape of the side slope at a location on the landscape. Traditionally, slope profile means the shape of the slope along its length when viewed longitudinally. It can be computed as an attribute value for each cell according to the value of that cell relative to the respective values of two *diametrically* (完全的) opposing neighbors. For analysis, it can be subdivided into 3 simple slope units: *straight*, *convex* and *concave units* (直线坡，凸坡，凹坡). A straight unit is called a *rectilinear segment* (直线段) characterized by a constant slope angle. On a slope, the maximum segment is the part that is steeper than the slope units above or below. Below the maximum segment is the concave slope and above it is a convex slope. Most likely they are found at different segments of the same slope.

More complex slopeshave several *manifestations* (表现形式). Each of them reflects a distinctive geomorphological process. Two popular examples of slope models are listed in Figure 9-6 and Table 9-3.

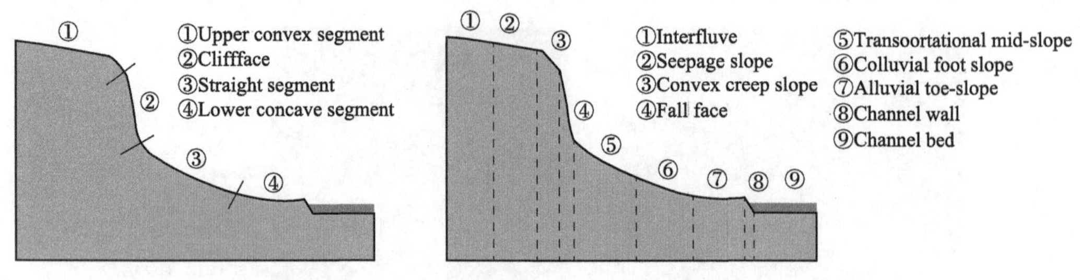

Figure 9-6 Slope Profile Models

Table 9-3 9-unit Slope Model According to Dalrymple in 1968

Classification	Unit	Slope	Slope process and characteristics
Waxing slope (秃顶峰)	Interfluve (河间地)	0°-1°	Soil forming process takes place; divide area characterized by largely vertical subsurface water and soil movement
	Seepage slope (渗流坡)	2°-4°	Chemical & mechanical movement of the material moving particles downslope; dominant Seepage of water
	Convex creep slope (凸蠕变边坡)	5°-20°	Hillslope dominant; soil creep occurring
Free face (自由面)	Fall face (崖壁)	45°-65°	Cliff face characterized by rapid*detachment* (脱离) of material or *bedrock* (岩床) exposure
Constant slope (直坡)	Transportation mid-slope (输送坡)	26°-35°	Active region characterized by the mass movement, *terracotta* (小土滑阶坎，阶地) formation, slope wash, and subsurface water action
Waning slope (减弱破)	Colluvial slope (崩积坡)		*Depositional region* (沉积区); material is further transported downslope by creep, slopewash, and subsurface flow
	Alluvial toe-slope (冲积坡)	0°-4°	Region of alluvial deposition
Humid temperate climatic regions (湿润的温带气候地区)	Channel Wall (河床壁)		Removal by *corrosion* (侵蚀), *slumping* (滑塌), fall, etc.
	Channel Bed (河床)		Transportation of material down the Valley Hillslope process unit

Based on *semi-arid* (半干旱) conditions, the 4-unit slope model (basic slope model) was put forward by *A. Wood* in 1942. The 4-unit slope model is usually developed on a high initial (primary) slope composed of strong rock and the absence of local undercutting. As the steep unit retreats, the base is covered by a *straight talus slope* (直堆积体边坡).

9.4.2 Specific Geomorphometry

Techniques for specificgeomorphometry have mainly focused on the delineation of terrain features related to surface hydrology. A range of features may be extracted from DTMs: surface-specific points (pits, peaks, passes, etc.), linear features (drainage channels and ridges), and areal features (drainage basins and hills). The objective of the analytical extraction process is to delineate the geometry of hydrological features, topologically connect them into contiguous networks, and obtain descriptive attributes for individual elements. This wealth of information may be used in applications such as *hydrological runoff simulation* (水文径流模拟), geomorphological modeling, support of interpolation procedures, or in cartographic generalization of DTMs.

9.4.2.1 Drainage Network Delineation(水系网络划分)

(1) Flow Direction

Water is assumed to flow from each cell to its steepest downslope neighbor; and if no neighbor is lower, the cell is a "pit" or "basin" and gets code 0. Assume only 8 possible directions of flow (up, up-right, right, down-right, down, down-left, left, and up-left) for each location; Number the move directions clockwise from up (Figure 9-7).

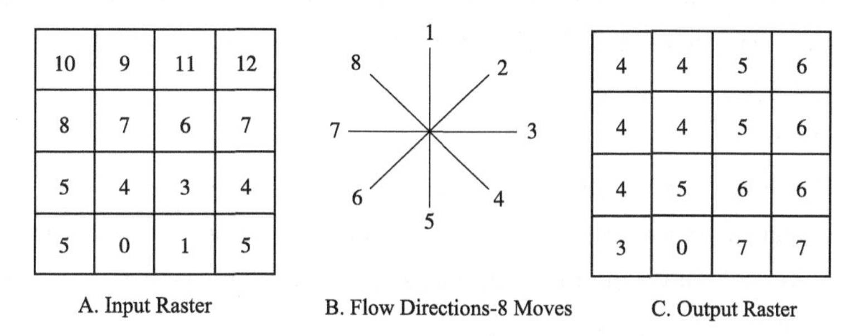

Figure 9-7 Flow Directions Sketch

(2) Flow Accumulation

The flow direction operation determines the natural drainage direction for every pixel in a Digital Elevation Model (DEM). Based on the output Flow direction map, the Flow accumulation operation counts the total number of pixels that will drain into *outlets* (排水口). Flow accumulation creates a raster of accumulated flow to each cell (Figure 9-8B).

(3) Flow Length

The function calculates the upstream or downstream distance, or weighted distance, along the flow path for each cell. The value type for the Flow Length output raster is floating-point. The

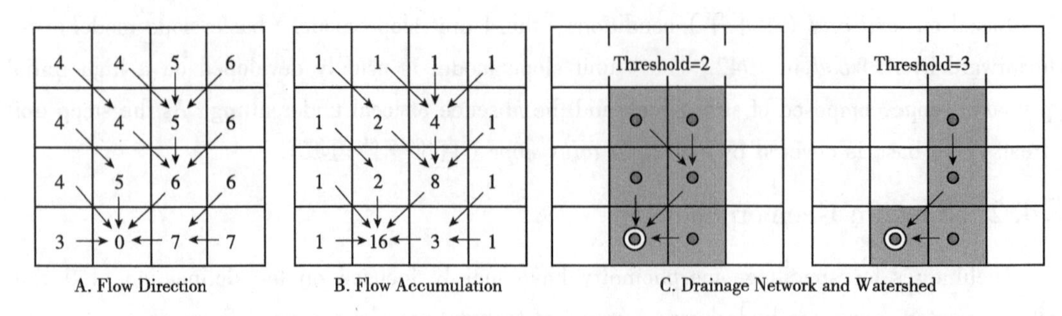

Figure 9-8 Derivation of the Watershed-Determines the Contributing Area Above a Set of Cells in a Raster

primary use of the Flow Length tool is to calculate the length of the longest flow path within a given basin. This measure is often used to calculate the time of concentration of a basin. This would be done using the UPSTREAM option. The tool can also be used to create distance-area diagrams of hypothetical rainfall and runoff events using the weight raster as an *impedance* (阻抗) to movement downslope.

(4) Drainage Network

A DTM contains sufficient information to determine general patterns of drainage and watersheds. Drainage modeling identifies locations located along the steepest downhill path extending from a "target area". The algorithm always chooses the direction of the maximum relief and search stops at the location where no more "downhill" neighbors can be found.

Drainage Network connects moves (of flow direction) with arrows. A zero on the edge of the array is interpreted as a channel that flows off the area. In most cases, the channel will not be formed until the accumulated water flowing down through the cells exceeds a *threshold volume* (阈值) (Figure 9-8C).

(5) Watershed

A watershed is defined as an attribute of each point on the network which identifies the region upstream of that point. Begin at the specified point and label all cells which drain to it, then all which drain to those, etc. until the upstream limits of the basin are defined. The watershed is the polygon formed by the labeled cells.

Example of GIS Drainage Network Delineation is shown as Figure 9-9.

9.4.2.2 Viewshed & Intervisibility

Predicting whether one point is visible from another (intervisibility analysis, 可视性分析) and predicting the total area which is visible from a single point (viewshed analysis, 可视域分析) are standard tools in GIS software. Viewshed analysis is mainly used to process the terrain in an optimum way, for instance, setting the radar station and the television transmission station, road construction, *maritime navigation* (海上航行), etc.

This function of intervisibility analysis computes an attribute value for each cell according to the visibility of that cell from one or more "viewer points". This often relates to data layers of ele-

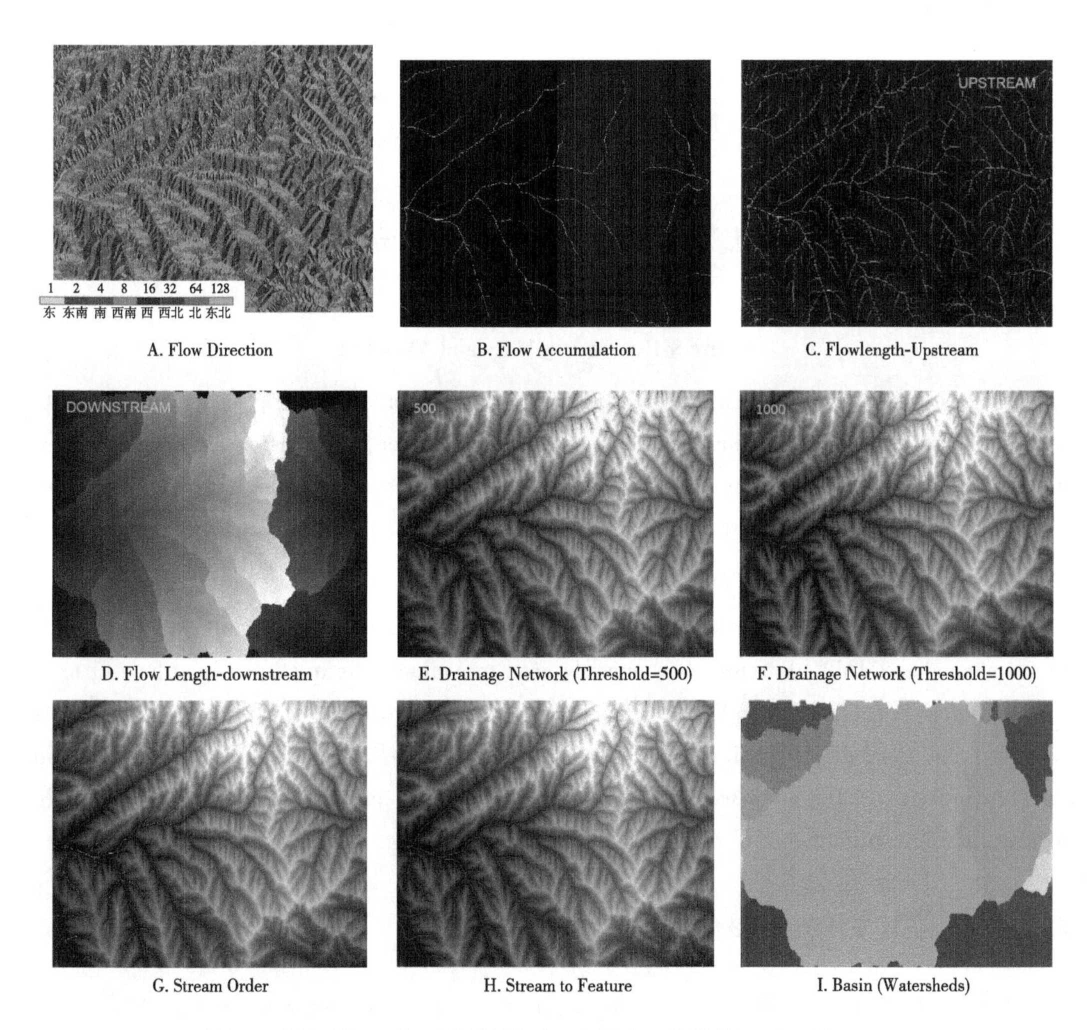

A. Flow Direction B. Flow Accumulation C. Flowlength-Upstream

D. Flow Length-downstream E. Drainage Network (Threshold=500) F. Drainage Network (Threshold=1000)

G. Stream Order H. Stream to Feature I. Basin (Watersheds)

Figure 9-9 Example of GIS Drainage Network Delineation Process

vation, viewer locations, and objects on the landscape which may influence the *line-of-sight* (视线) between viewers and locations on the landscape. If the land surface rises above the line of sight, the target object is invisible. Otherwise, it is visible from the viewpoint. The line-of-sight computation is repeated for all target pixels from a set of viewpoints and the set of targets that are visible from the viewpoints from the viewshed (Figure 9-10).

9.4.3 Interpretation for DTM Quality Assessment

9.4.3.1 Error Detection and Correction

Errors-*blunders* (错误) and constant, systematic, and random errors-may occur in any sampling process. Apart from geometrical errors, classification errors may occur (e. g. an edge in a DTM may be classified as part of breaking when it is not). Procedures for detection and correction of errors are thus important.

The method most often used is *visual inspection* (目视检查) and interactive editing of the incor-

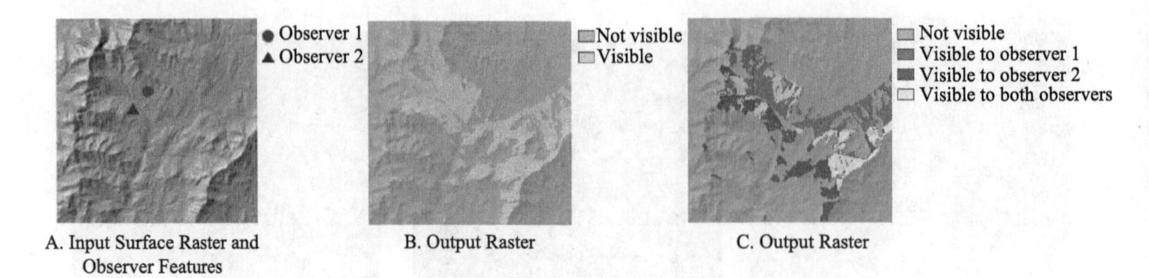

A. Input Surface Raster and Observer Features B. Output Raster C. Output Raster

Figure 9-10 Intervisibility and Viewshed

rect elements. Some display techniques are particularly useful for highlighting errors: on perspective displays, errors are detectable since they project out of the surface, whereas on hill shading or maps of slope (or of other geomorphometric parameters) errors become obvious due to gradient *anomalies* (异常). Apart from gradients, special error indices may be computed and visualized.

9.4.3.2 DTM Quality Control

Quality control of a DTM may be performed by comparison with reference data (i. e. control points or another DTM). Parameters commonly used to evaluate the quality of a surface fit to reference points (e. g. Root Mean Square Error, RMSE), are reviewed by *Willmott* (1984). Other techniques are used for comparing two DTMs of the same area: statistical analysis of residual surfaces, comparative analysis of *semi-variograms* (半方差函数) or *frequency spectra* (频谱). Procedures for DTM quality control may also be used to help detect systematic or constant errors.

9.5 DTM Visualization

Results of DTM modeling operations are most often communicated to users in graphical form. Visualization thus plays a vital role in a DTM system. It is closely linked to interpretation: results of interpretation steps need to be displayed, and interpretation operations may in turn lead to improvements in visualization. Moreover, graphics themselves may directly support decision-making (through visual interpretation) without involving any quantitative analysis.

Visualization commonly pursues two goals: interactive visualization, which helps the researcher to explore models and refine *hypotheses* (假设); and static visualization (in the traditional form of paper maps), which is used to communicate results and concepts. The usefulness of visualization products mainly depends on their communicational effectiveness and their ability to support interpretation. In this light, utmost realism is not a primary objective.

9.5.1 Contouring

Contour lines (isolines) are probably still the most widely used technique for displaying relief. Contours represent a method for quantitative visualization of 3D. Contour displays or hill shading may be overlaid with other elements such as the results of interpretation procedures, components of the DTM, or other 2D data. They are used to satisfy the requirement for extracting quantitative information from relief displays (e. g. in geology, topographic mapping, and civil engineering). The major drawback of contours is that they give no immediate impression of the topographic forms.

The construction of contours is closely related to DTM interpolation. Basic contouring may be further refined: *index contours* (计曲线) may be highlighted by special symbolization (Figure 9-11); Contour labels may be added for index contours, and contour drawing may be suppressed in areas of steep gradient (i. e. where contours are too densely *cluttered* (混乱), only index contours are drawn).

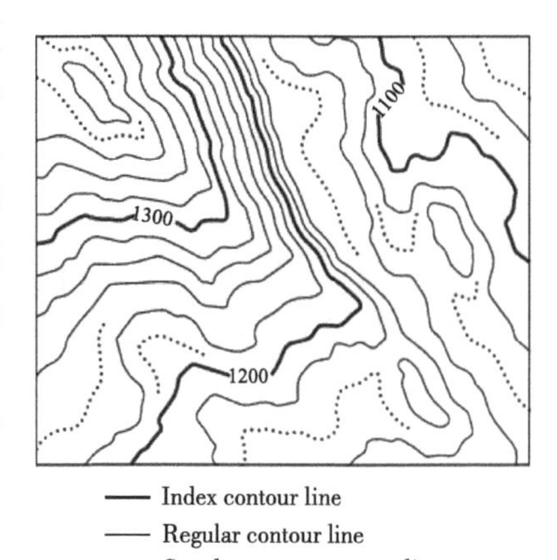

Figure 9-11 Contour Lines

Several alternative methods exist for contour line display. Elevation data may simply be classified into *equidistant altitude classes* (等距高程分类) to produce a *hypsometric tint display* (分层设色显示). Alternatively, gridded DTMs may be contoured by a technique called "raster contouring": instead of threading contour lines through the grid, contours are determined based on individual grid cells. Finally, two methods-called "inclined contours" and "relief or shaded contours"-may be used. Both techniques attempt to extend contouring to give a better 3-dimensional impression of relief. The first method represents intersections of the surface with parallel inclined planes that are *orthogonal* (直角正交的) to the direction of the light source (i. e. inclined contour lines). The latter method draws contours by varying their width according to illumination brightness.

9.5.2 Hillshading

Landforms may be readily perceived from shaded relief displays. *Hillshading* (地貌晕渲, relief shading) is an important technique developed by cartographers to improve the visual qualities of maps. It provides a convenient way of qualitative cartographic relief depiction. The method is based on a model of illumination. There are several models available to produce relief shade using DTM. The most commonly used model is based on the *optical theory* (Lambert's Cosine Law, 余弦定理) which states that the brightness of any small area of a perfectly *diffuse undulating surface* (弥散起伏的地表) varies as the cosine of the angle of incident parallel light. Calculation of percent intensity of reflected sun light (太阳光反射光线强度的百分比):

$$I=\frac{100}{c}\left(\sin\phi-\frac{\delta x}{\delta z}\sin\alpha\cos\phi-\frac{\delta z}{\delta y}\cos\alpha\cos\phi\right) \tag{9-2}$$

$$c=\sqrt{1+\left(\frac{\delta z}{\delta x}\right)^2+\left(\frac{\delta z}{\delta y}\right)^2} \tag{9-3}$$

where I is the amount of sunlight reflected by a surface element; α and ϕ denote *azimuth* (方位角) and elevation of the sun, respectively (Figure 9-12); c is the *scalar* (标量).

Figure 9-12 Hillshading Using a DTM

There are many different shading algorithms, the most well know are *flat shading* (平面晕渲) and *smooth shading* (平滑晕渲). The key difference between flat and smooth shading is in the way that the normals are used. Flat shading simply assigns each triangle a normal vector and the lighting is done individually on each face. For smooth shading, the surface normals of the surrounding faces are averaged and at each vertex is assigned a normal. Flat (or constant) shading is valid for small objects and if the source light and the viewer are at infinity; for high detail levels, a great number of flat-shaded polygons is required, therefore it is of little value for realism. Smooth (or interpolated) shading can be applied with many algorithms, but the two "classic" approaches are Gouraud and Phong. *Gouraud shading* (高洛德着色、高氏着色、高氏渲染) specifies a color for each vertex and polygon and then intermediate colors are generated along each edge by interpolation between the vertices. *Phong shading* (补色渲染) requires a normal interpolation for each pixel and so is quite prohibitive and time consuming for real-time use.

9.5.3 Combination with 2D Data, Orthophotos

Contour displays, as well as hill shading, may be overlaid with other elements such as the results of interpretation procedures (e. g. visibility, slope, and drainage networks); components of the DTM (e. g. data points, structural features, and TIN triangles); or other 2D data (e. g. roads, land use, and geological maps). To combine areal data (e. g. land use) with hill shading, the color or intensity values of the areal data have to be *modulated* (被调整的) with the shading intensity at the corresponding location.

Orthophotos (正射影像) is a further orthographic display technique. Orthographic displays provide the advantage that all parts of the terrain surface are visible and relatively undistorted. They are generated from overlapping conventional aerial photos in a process called *differential rectification* (微分纠正) (using DTMs) to eliminate image distortions due to topography. Therefore, Orthophotograph can be considered as a modified copy of a perspective photograph of the earth's surface with distortions due to tilt and relief removed. Orthophotos are commonly used in planning applications or topographic mapping.

9.5.4 Perspective Display

The advantage of orthographic displays is that all parts of the terrain surface are visible and relatively undistorted (Figure 9-13). *Perspective displays* (透视显示), on the other hand, provide much more convincing visualization results for DTM. It is common to display a DTM in a perspective view in GIS applications. Two of the main problems that have to be solved for perspective display are the projection of the 3D surface onto a 2D medium and the elimination of hidden elements from the display. The 3D and perspective transformations (including 3D clipping) necessary for perspective projection are described in standard textbooks on computer graphics.

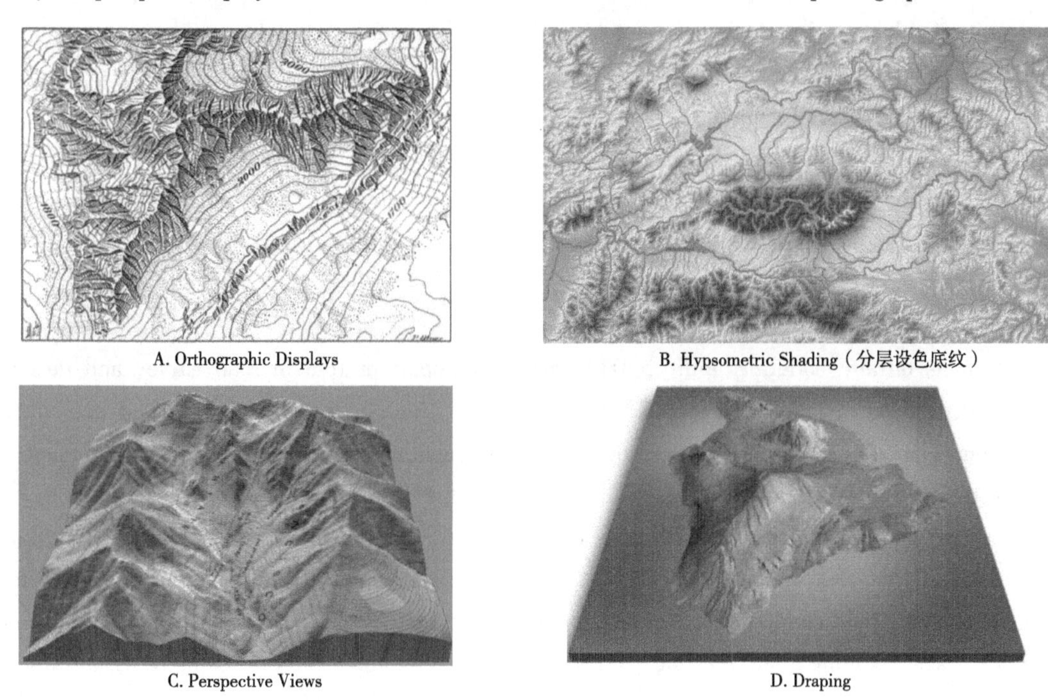

A. Orthographic Displays　　B. Hypsometric Shading (分层设色底纹)

C. Perspective Views　　D. Draping

Figure 9-13 DTM Visualization

(Draping: A perspective or panoramic rendering of a 2D image superimposed on a 2D-or 2.5D surface. Drapes are usually, but not necessarily, remotely sensed images used to provide context to a Digital Terrain Model)

Panoramic views (全景显示) are a special *variant* (变体) of perspective displays: The observer is located on or near the surface, and most often the view extends over a full circle.

9.6 DTM Application

DTM is closely related to many disciplines in theory and technology. From the point of view of data acquisition, information extraction, and space object expression, DTM is closely related to surveying and mapping, photogrammetry, cartography, remote sensing, geographic information system, and so on. It is an important research content of modern geographic information science.

From the point of view of data interpolation, data management, information processing, spatial analysis, and digital product output, DTM originates from the interdisciplinary intersection of traditional surveying and mapping and information science. In terms of theory, method, and technology, DTM draws on the relevant internal and latest development of numerical analysis, discrete mathematics, signal processing, database, graphics, and imaging.

At the application level, as a new generation of a topographic map, DTM is widely used in various industries and fields and has become the basic data and powerful tool to support the application of geoscience modeling and spatial analysis. At the same time, as the basic product and key technology of the geographic information industry, DTM is constantly exploring new application fields, playing an important role in industrialization and social services.

At present, the application of DTM mainly covers *civil engineering* (土木工程, transportation, water conservancy, mining), regional urban and rural planning (regional planning, urban design, scenic area planning), geoscience research, natural resources investigation and management, environmental monitoring and evaluation, disaster mapping and emergency support, military defense, and other fields. The following is a brief description of four topics.

9.6.1 Civil Construction and Other Works

With its terrain modeling ability, DTM plays an important role in road survey and design (highway, railway), geological exploration (oil, coal, etc.), mine surveying and mapping, water conservancy engineering construction, telecommunication, and power engineering projects.

In the road survey and design, the designers use the road CAD system to collect the topographic data along the section and use DTM to carry out the alignment analysis and route design of the horizontal, vertical, and cross-sections. It can also be used to carry out *the earthwork calculation* (土方计算) and economic evaluation, compare and analyze the road route, and through the terrain surface modeling along the line, DTM overlay analysis and road landscape 3D visualization can formulate the scientific and economic engineering design scheme for road construction.

In civil engineering and other engineering software, the functions of real-time generation, graphic editing, professional analysis, and 3D visualization of DTM are highly integrated. They not only support road design but also are widely used in municipal pipe networks, oil pipelines, hydraulic channels, telecommunication power line design, and other engineering fields.

In geological exploration and mine surveying and mapping, DTM, geological map, remote sensing image, combined with GIS spatial analysis can construct a three-dimensional surface landscape, and it can be used to carry out *geological section survey* (地质剖面调查), *surface subsidence* (地面沉降) and deformation monitoring, *waste mine land reclamation* (废弃矿区复垦) and ecological reconstruction, mining area 3D simulation system development and so on, which is of great significance to understand the complex surface and underground three-dimensional geological environment.

In the water conservancy project (e. g. river channel or reservoir area construction), DTM combined with the hydrological observation data and remote sensing images, 3D terrain modeling and *reservoir capacity* (水库库容) can be developed to calculate flood simulation, dam site selection analysis. Sometimes, DTM can also provide technical and data support for hydrological monitoring, including generating and drawing *isoline map* (等值线图) of reservoir water level, water level area curve, river section map, etc.

9.6.2 Digital City and Urban/Rural Planning

DTM plays an important role in regional planning, urban design, and *scenic area planning* (景区规划). It is a comprehensive technical system that can be used to support urban/rural planning decision-making, implementation, and evaluation. Combined with remote sensing, GIS, and survey data, DTM plays a supporting role in three-dimensional visualization expression and virtual reality representation.

Taking urban planning as an example, topography and geomorphology are the basic elements that restrict the spatial layout of urban land, regional ecological pattern, visibility corridor, and spatial landscape elements. Especially in mountainous cities, when making regulatory detailed planning, elevation, slope, aspect, and other issues must be considered when considering the land layout. At present, paper topographic maps have been replaced by GIS, CAD, and other information support environments as the base map to format urban architectural space. The 3D visualization geographic environment based on DTM can provide comprehensive support for urban/rural planning and management.

More importantly, as the rapid development of real 3D city construction in recent years, it mainly uses *airborne lidar* (机载激光雷达), *digital measurement camera and tilt camera* (倾斜摄影相机) to obtain high-precision terrain, top and side images of the city, quickly and actively generate large-scale DTM, DOM, DLG, model texture, tilts real image and other "digital city" products.

9.6.3 Earth Science and Emergency Support and Mapping

DTM is the basic data source for geoscience research, such as geography, geology, hydrology, environment, atmosphere, and so on. At the same time, combined with remote sensing, it also plays an important role in emergency support and mapping.

In hydrology research, the distributed hydrological model can be constructed by using DTM, thematic map and remote sensing data, combined with *meteorological* (气象的), *hydrological* (水文的), soil, remote sensing, surface field observation data. *The distributed hydrological model* (分布式水文模型) is a quantitative hydrological model based on *the watershed units* (流域单元) divided by DTM. DTM can be used to provide the morphological parameters of the land surface, including the slope, aspect, and the relationship between the elements. According to the algorithm, the surface water flow path, river network, and basin boundary can be calculated, and then the "soil-vegetation-atmosphere" system can be calculated in the unit.

The development of *digital geomorphology* (数字地貌) is an important trend of modern geomorphology. The basis of geomorphology research is to describe the morphological characteristics of surface topography parameters, including the quantitative expression of elevation, slope, aspect, topographic relief, and other parameters of the study area. The spatial analysis and digital image processing technology based on DTM can quickly extract the characteristic parameters (i. e. basic geomorphic factors) of landforms. It can also further analyze complex geomorphic phenomena by using DTM spatial analysis, geomorphic classification and mapping, constructing a landscape geomorphic model, and conducting quantitative analysis on its topographic profile.

In the study of *meteorology and climatology* (气象学和气候学), combined with DTM analysis, the local terrain environment can be underlaid. The impact of surface conditions (mainly topography and geomorphology) on regional climate is evaluated, such as elevation distribution, variation of *solar radiation* (太阳辐射) (altitude angle, 仰角大小; shadow area, 阴影区域; radiation intensity, 辐射强度), surface water and groundwater movement, surface wind change and airflow movement, terrain slope, geomorphic enclosure conditions and so on. These factors need to be quantitatively analyzed according to DTM. At the same time, terrain analysis is also needed when setting up *meteorological observation stations* (气象观测站). The elevation, slope, aspect, and visibility conditions are the evaluation factors, which can help to select the better station location.

When major disasters, accidents, and disease outbreaks occur, emergency surveying and mapping support services need to be carried out. DTM plays an irreplaceable role in monitoring and disaster relief of geological disasters (earthquakes; *volcanoes*, 火山; *landslides*, 滑坡) and secondary natural disasters.

9. 6. 4 Military Applications

National defense construction and military operations are the most important users of DTM products, and they are also the most important users of DTM products.

In the traditional war before modern times, military operations have high requirements for the accuracy of terrain. All aspects of *the battlefield environment* (战场环境) are closely related to the topography. Campaign organization often needs to make a comprehensive analysis of battlefield terrain through a paper topographic map or *terrain sand table* (地形沙盘) combined with field battlefield *reconnaissance* (战场侦察) to accurately grasp terrain elements such as landform, water system, road, residential area, soil and vegetation, and to judge their influence on army movement, observation, shooting, *concealment* (隐蔽) and *camouflage* (伪装). In combat simulation and wartime command decision-making, senior commanders, with the cooperation of staff members, draw or manually move targets on the headquarters map to depict or deduce the battlefield situation in real-time.

While in nowadays DTM is the foundation of "digital battlefield" construction and the core of military surveying and mapping support data products. In the process of fighting, the front-line commanders need to use the favorable terrain characteristics derived from DTM to build fortifications and carry out *force allocation* (兵力部署). By setting and organizing the march route, the commanders can achieve the purpose of seeking advantages and avoiding disadvantages by relying on the terrain correctly.

Vocabulary

algorithms[ˈælgərɪðəm] *n.* 算法
aspect[ˈæspekt] *n.* 方面；方位；外观
bare earth 裸露地表
cartographic[ˌkɑːtəˈgræfɪk] *adj.* 地图的；制图的
circumcircle[ˈsɜːkəmˈsɜːkl] *n.* 外接圆
criteria[kraɪˈtɪəriə] *n.* 标准；尺度；准则
cross-section[ˈkrɒsˈsekʃən] *n.* 横断面；截面图
Delaunay triangulation/Delaunay 三角剖分
delineation[dɪˌlɪniˈeɪʃn] *n.* 画轮廓；略图；图形
derivation[ˌderɪˈveɪʃn] *n.* 派生；推导；来历；衍生物；导数
drainage networks *n.* 排水网络
DTMs 数字地形模型
geomorphometry[gjoʊ mɔːˈfɔmɪtrɪ] *n.* 地貌量计学
hypsometric[hɪpsəˈmetrɪk] *adj.* 测高术的
illumination[ɪˌluːmɪˈneɪʃn] *n.* 照明；阐释；启发；古书上的图案或装饰
index contours 计曲线
instruments[ˈɪnstrəmənt] *n.* 仪器；乐器；工具；器械
interpretation[ɪnˌtɜːrprɪˈteɪʃn] *n.* 解释；翻译；阐释
isolines *n.* 等值线(isoline 的复数形式)；网线密度
line-of-sight[ˈlaɪnˈəvsˈaɪt] *n.* 视线
manipulation[məˌnɪpjuˈleɪʃn] *n.* 操纵；控制
non-periodic[ˈnɒnpiərɪˈɒdɪk] *adj.* 非周期的
non-stationary[ˈnɒnstˈeɪʃənrɪ] *adj.* 非平稳；不稳定的
normal vector 法向矢量
orthophotos[ˈɔːθəʊfəʊtəʊz] *n.* 正射影像
perspective display[pərˈspektɪv] 透视显示
phenomena[fəˈnɒmɪnə] *n.* 现象
photogrammetric[foʊtoʊgrəˈmetrɪk] *adj.* 航空摄影测量的
profile[ˈproʊfaɪl] *n.* 剖面
relief shading 地貌晕渲
resampling[reˈsɑːmplɪŋ] *n.* 重采样
simulation[ˌsɪmjuˈleɪʃn] *n.* 模拟；仿真；赝品

subaquatic[ˌsʌbəˈkwætɪk] *adj.* 半水生的
subsequent[ˈsʌbsɪkwənt] *adj.* 随后的；后来的
TIN 不规则三角网
topographic maps 地形图
visualization[ˌvɪʒuəlaɪˈzeɪʃn] *n.* 可视化；形象化

Questions for Further Study

1. What are digital terrain models(DTMs), and what are their characteristics?
2. What are the main tasks of DTMs, and please describe each section of these tasks?
3. Which do DTMs data structures include?
4. What are the characteristics of gird DTMs data structure, and what is that of TINs?
5. Which are DTMs' data sources, and what are their characters?
6. Which methods do terrain visualization include, and what are the characteristics of each method?
7. What is the base of the slope factor extraction algorithm?
8. How can slope and aspect factors be extracted by using GIS?
9. What principle is drainage network extraction based on grid DEM, and please draw their technical route?

Chapter 10

Network Analysis

This chapter is one of a series of spatial functions analysis about Vector GIS discussed in Chapter 8. In this chapter, the discussion is more closely related with network.

10.1 Definition of Network

Networks are all around us. Roads, railways, cables, pipelines, streams and even glaciers are phenomena that frequently need to be represented and analyzed as a network. They are used to transport people and goods, communicate information and control the flow of matter and energy. Most of the economic and social activities in the world are organized into networks. The form, capacity and efficiency of these networks have substantial impacts on our living circumstance and human perception of the world around us. Network analysis enables people to solve problems, such as planning travel destinations and services based on travel time, choosing the most efficient travel route and the closest public facilities (网络无时无刻不在我们身边。公路、铁路、电缆、管道、溪流甚至是冰川，经常需要用网络来表达和分析。网络常被用于人流移动、货物运输、信息交流以及物质控制和能量的流动。世界上大多数的经济和社会活动都可利用网络来进行组织。庞大的网络信息和效率深刻影响着我们的生活环境及对周围世界的看法。网络分析使我们能够解决很多问题，如规划最佳的旅游路线、寻找旅游目的地、找到最近的公共设施、根据时间规划目的地出行)(Figure 10-1).

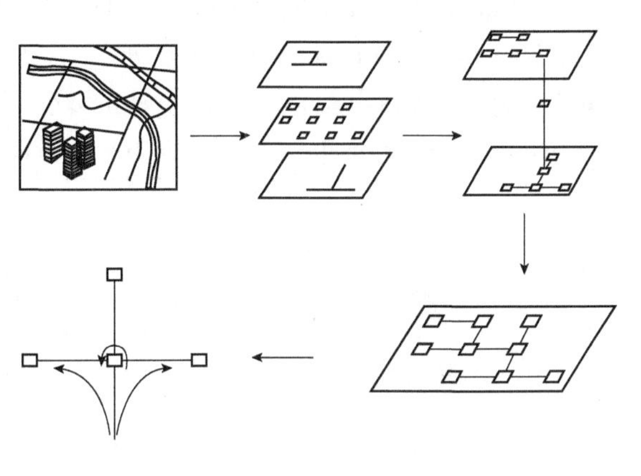

Figure 10-1 Network Construction and Analysis

Network is an interconnected set of arcs or lines representing possible paths for the movement of resources from one location to another, or a group of computers linked and enabled to share *peripherals* (周边设备), software and data. A network data model is based upon the idea of *explicit* (显式) links between

Figure 10-2　Examples of Network

related entities. The Internet is probably the most well-known example of a computer network and a type of database structure (Figure 10-2).

10. 2　Basic Elements

10. 2. 1　Links and Nodes

Network links are interconnected linear entities which represent the *conduits* (管道) for transportation (e. g. vehicles, fluids, electricity) and communication networks. For example, highways, electrical transmission lines and the *ethernet cables* (以太网电缆) that form a computer network are all network links. Network link attributes describe individual link characteristics, such as one-way or two-way lanes, etc. In the GIS literature, a network link is often called a network arc.

The difficulty of moving from one node to another in a network link is the link *impedance* (阻抗). Sometimes the impedance means *total resistance* (总阻力). For modeling, link impedance can be expressed as distance, but travel time or apparent cost is usually better measures. The link impedance depends on the direction of travel (Figure 10-3).

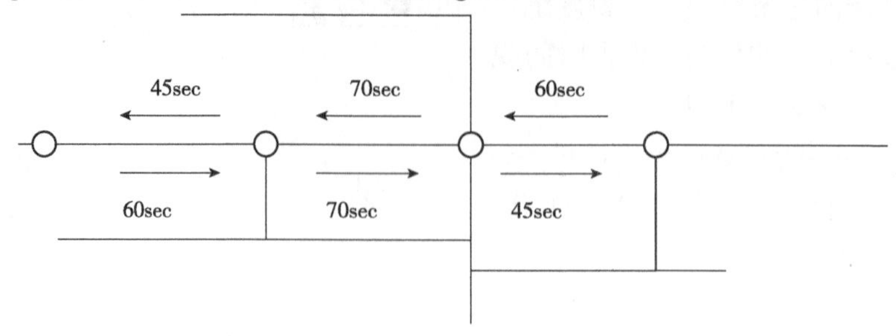

Figure 10-3　Network links

As the endpoints, network nodes connect points ofnetwork links, such as *intersections*（十字路口）and *interchanges*（立体交叉道）of a road network, the confluence of streams in a hydrologic network, and switches in a power grid. In computer networks, a network node is a computer attached to that network. Local nodes or remote nodes in a network may all offer a set of barrier counters. Network barriers block or *impede*（妨碍，阻止）something in a network and sometimes form a concurrent *synchronization*（同步）via the *on-chip*（片上）network.

10.2.2 Edges, Junctions and Turns

Generated from the source features, network datasets are composed of network elements which help establish connectivity. In addition, network elements have attributes that control *navigation*（导航）over the network. They include the following three types:

①Edges: Connecting to other elements (junctions), they are the links over which agents travel.

②Junctions: Connecting edges and facilitating navigation from one edge to another.

③Turns: Storing information that can affect movements between two or more edges. A turn on a network is the transition from one arc to another arc on a network.

Edges and junctions form the basic structure of a network. Connectivity in a network deals with connecting edges and junctions to each other.

Network turns represent movement relationships between network links; they are created to increase the cost of making the movement, or prohibit the turn entirely. For instance, a right turn against on-coming traffic takes more time than straight forward traffic; a left turn at an intersection takes 30 seconds, which is the average time it takes for the left turn light to turn green. Similarly, a restriction attribute could read a field value from a turn feature to prohibit it. This is useful when the turning movement is posted as illegal (no left turns).

Turns can be created at any junction where edges connect. There are n^2 possible turns at every network junction, where n is the number of edges connected at that junction. Even at a junction with a single edge, it is possible to make one U-turn (Figure 10-4).

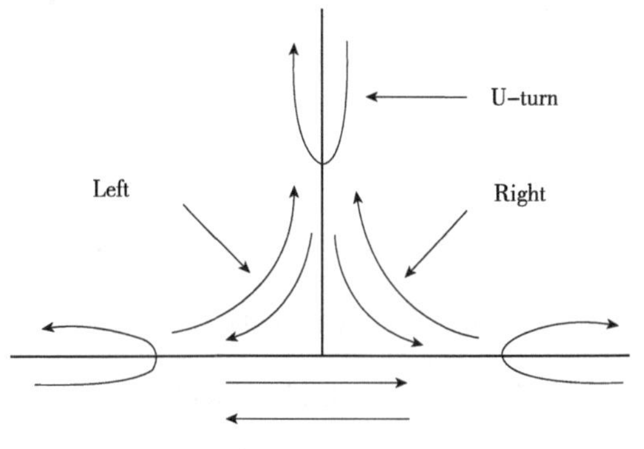

Figure 10-4 Example of Turns

10.2.3 Paths, Tours, Stops and Centers

A path is an alternating sequence of nodes and links, beginning and ending with nodes. Repeating nodes and links within a path are permitted, but are rare in most network applications. A shortest path is the shortest or most cost-saving path from a source node (origin) to a destination node. In practice, *path finding*（领航，寻找目标）is to find the shortest or most efficient way to

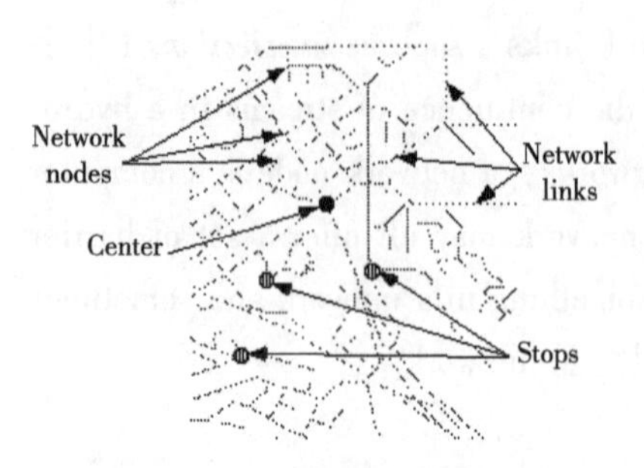

Figure 10-5 The Basic Elements of Network

visit a sequence of locations.

A tour is similar to a path, but it is a closed one. It means that the first node and the final node on the path are the same on the network.

A stop is a location visited in a path or a tour.

A center is a location where certain resources are supplied.

The basic elements of a network are shown as Figure 10-5.

10.3 Network Data Structure

10.3.1 Types of Networks

A network can be directed or undirected, although links and paths typically have direction.

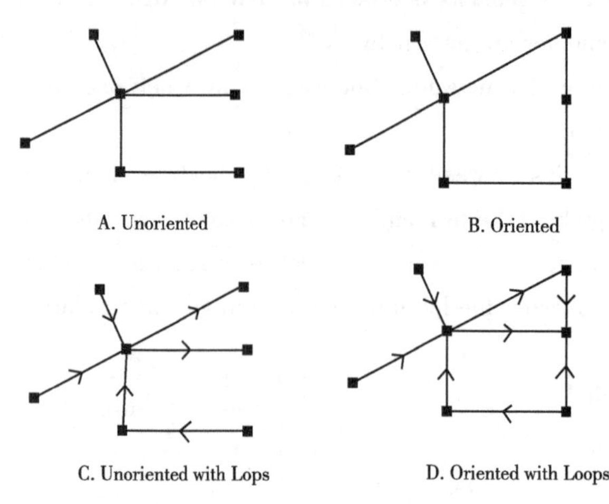

Figure 10-6 Types of Networks

A network can be directed or undirected, although links and paths are usually directional. A network is directed if the links it contains are directed, and undirected if the links it contains are undirected.

So, types of network can be divided into: unoriented network and oriented network, both the two types of network with *loops* (环形线路) are as shown in Figure 10-6.

For example, river network and gas pipeline network should be oriented; airline network may be unoriented; railway networks and road networks may be either un-oriented with loops or oriented with loops.

10.3.2 Topological Classification of Networks

Formally, networks are digitally represented by nodes and links(一个网络可由节点和链来进行数字化表达). In this relationship, nodes represent intersections, interchanges and confluence points, while links represent transportation facility segments between nodes. Among several types of networks in computer technology nowadays, the two major topological ones are planar network and non-planar network. In a planar network, no links intersect except at nodes such as road and highway networks. Conversely, links may intersect anywhere in the non-planar network such as airline networks (Figure 10-7).

An overpass of one road over another road in a planar model is generally represented by a node with four incident arcs. The node at the intersection has an associated attribute describing turn restrictions in order to convey the correct driver instructions (i. e. for route guidance). In a non-planar model of the same feature, no node exists since the arcs are unconnected in the 3-dimensional space. Thus, linking non-planar and planar databases is problematic.

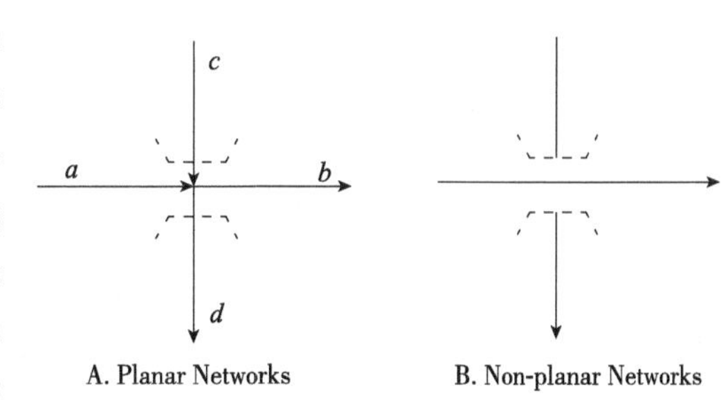

Figure 10-7 Topological Classification of Networks

Most networks are entered as linear graphs in a vector GIS, with attribute data recorded with links rather than nodes. Examples for link description include traffic load, capacity of the road segment, etc. However, the link-based approach is not entirely suitable for applications that deal directly with nodes in a network. A typical example is the *air passenger volume* (航空客运总量) between cities where there is no direct link (multiple routes).

10. 3. 3 Network Level of Details

An essential attribute of the network model is strictly hierarchical. A link included in a lower-level network model should also be included in each higher-level network model. These hierarchical relationships could guarantee very consistent network model and thereby enable an easier tracing of the effects of all the levels in details.

Three network has been developed into three levels: fine, medium and coarse, which are illustrated in Figure 10-8. The fine level is nearly identical to the actual road network. It includes almost all streets and has building blocks as zones. The medium level includes all *arterials* (主干道) and collectors and corresponds with normal transportation planning practice. The coarse level represents only the arterials and may therefore be regarded as a sketch-planning network.

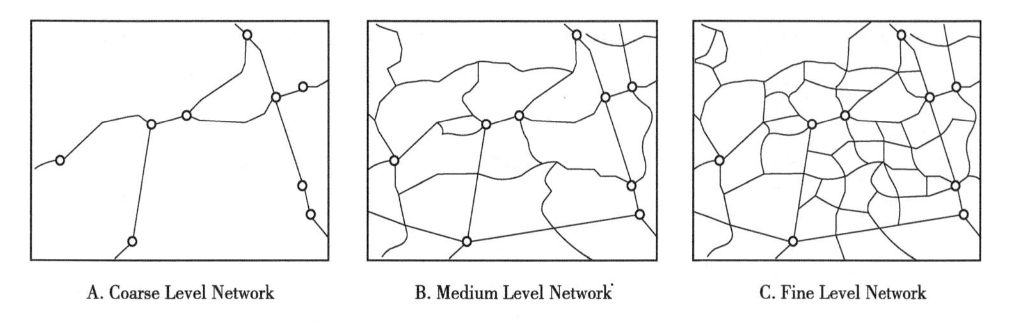

Figure 10-8 Network Level of Details

10.4 Common Analysis Capabilities

10.4.1 Routing or Flow Analysis

Routing (路径选择) includes "route" and "best". "Route" can be simple such as finding driving directions between two points; it is also complex such as finding the best route between 10 different stops. Here, "best" means such things as the shortest distance, etc.

Routing or flow analysis is to find the shortest, cheapest, quickest or optimal route between locations, or design the delivery program according to optimization criteria such as the least time, minimum cost, shortest distance, etc.

①The Shortest Path (for directed and undirected networks): A classical algorithmic problem in graph theory research. Given a start node and an end node, find the shortest path.

Finding the shortest route is probably the commonest routing problem for GIS users. Finding the shortest route from *A* to *B* through a road network is crucial for emergency services, business journeys, or simply holiday tours in a region. In order to carry out such operations, it is important to construct an appropriate network. Such details as connectivity, one-way streets, possible turns and speed limits will be considered. Existing shortest path algorithms can be categorized into two groups: label-setting and label-correcting. Both groups of algorithms are *iterative* (迭代) and both employ the labeling method in computing one-to-all (one node to all other nodes) shortest paths. These two groups of algorithms differ in the ways they update the estimate (i. e. upper bound) of the shortest path distance associated with each node at iteration and in the ways they converge to the final optimal one-to-all shortest path.

In label-setting algorithms, the final optimal shortest path distance from the source node to the destination node is determined once the destination node is scanned. Hence, if it is only necessary to compute a one-to-one shortest path, then a label-setting algorithm can be *terminated* (终止) as soon as the destination node is scanned. There is no need to exhaust all nodes on the entire network. In contrast, a label-correcting algorithm treats the shortest path distance estimates of all nodes as temporary and converges to the final one-to-all optimal shortest path distance until the shortest path from the source node to all other nodes is determined.

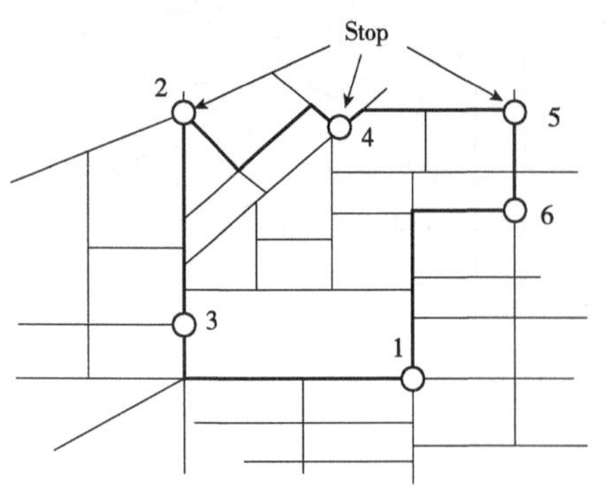

Figure 10-9 Example of TSP

②Minimum Cost Spanning Tree (for undirected networks): Given an undirected graph, find the minimum cost tree that connects all nodes.

③All Paths Between Two Nodes: Given two nodes, find all possible paths between them.

④Traveling Salesman Problem (TSP) Analysis: The TSP is a classical problem on combinatorial optimization, which is simple to state but very difficult to solve. Given a set of nodes, find the lowest cost path that visits all nodes exactly once and eventually returns to the initial node. The example of TSP is shown as Figure 10-9.

The travelling *salesperson* (推销员) has to return to the base after visiting a number of customers. He(She) needs to have details of the shortest tour possible to the visited sites. This is the obvious application for the shortest tour routes. It is possible for a travelling salesperson to traverse the same road twice. The travelling salesperson problem (TSP) has attracted the attention of operational researchers. Calculating the shortest distance between stations and then list all combinations of those stations that can be visited to find the shortest travel route. For each *permutation* (排列) you have to calculate the shortest route between successive places to visit and sum the results. Finally, you can compare the total distances travelled to identify the optimum solution.

10.4.2 Location and Allocation

Location-allocation (区位-配置) is the process of determining the optimal locations for a given number of facilities based on some criteria and simultaneously assigning the population to the facilities. Location-allocation analysis is commonly used in both the public and private sectors. The determination of locations for *retail stores* (零售店), restaurants, banks, factories, and warehouses is used in the private sector. In the public sector, the choice of locations for libraries, hospitals, post offices, and schools can be supported by analysis results from location-allocation models.

Location on networks involves selecting network locations on an existing network so that an objective is optimized. A few examples of location are as follows:

①Address Matching: Finding spatial locations based on address descriptions. Address can be interpreted to give approximate spatial location on a network.

②Within-cost Analysis (for directed and undirected networks): Given a target node and a cost, find all nodes that can be reached by the target node within the given cost.

③Nearest-neighbors Analysis (for directed and undirected networks): Given a target node and number of neighbors, find the neighbor nodes and their costs to reach the given target node.

Allocation is the process to assign portions of a network to a location (e. g. a center) based on some given criteria. The aim of resources allocation is to optimally locate a set of objects so that a variable or variables will achieve a maximum or minimum value. A few examples of allocation are as follows:

①Minimizing average distance.

②Minimizing the maximum distance to closest supply center.

③Minimizing the number of centers required for every demand point and keeping them within a critical distance of a supply point.

④Minimizing average distance subject to a maximum distance constraint.

Generally, "supply" means the amount of resources available for a facility (center); "demand" means the amount of resources demanded on the links or nodes. Allocation works on assigning demand usually to the nearest center until the demand matches the supply of the center. For example, centers may be schools with a maximum capacity for children, health centers with a capacity for patients, or warehouses with a capacity for goods (Figure 10-10 and Figure 10-11). Allocation algorithms use these centers as destinations and then model how people or goods will travel through the network to get there. Consequently, a map is used to show the areas served by each service facility e. g. a school or health center *catchment* (集水) area, or the warehouse's distribution area. The algorithms usually work on allocating links in the network to the nearest center, taking into account, of course, the attributes such as one way streets, barriers to movement and so on.

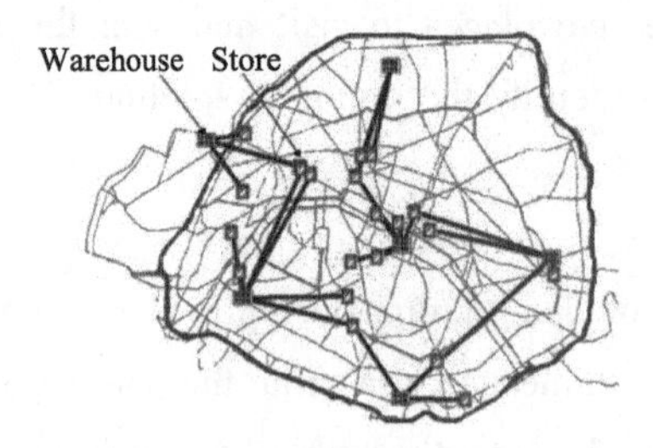

Figure 10-10 The Best Location of Goods

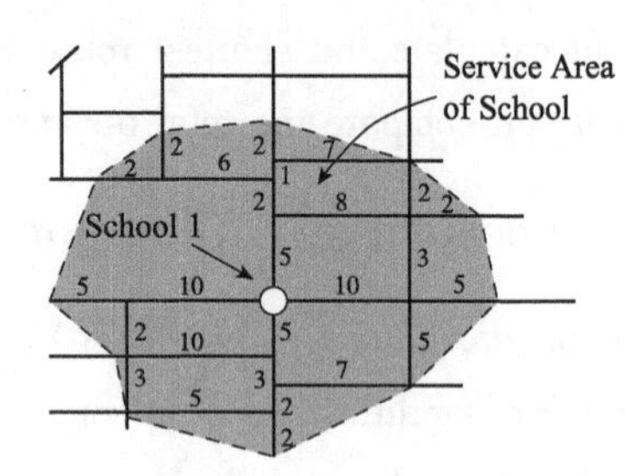

Figure 10-11 Allocation of School

10.4.3 Accessibility or Reachability

Accessibility or Reachability (可取性或可达性) provids an *aggregate measure* (总体措施) of how accessible a location is to other locations (Figure 10-12). It can be defined as the ease of participating in activities. The concept of most of the accessibilities is as follows:

①Aggregate Distance Minimization Criterion: The total of the distances of all people from their closest facility is minimum.

②Mini Max Distance Criterion: The farthest distance of people from their closest facility is minimum.

③Equal Assignment Criterion: The number of people in the proximal area surrounding each facility is approximately equal.

④Threshold Constraint Criterion: The number of people in the proximal area surrounding each facility is always greater than a specified number.

⑤Capacity Constraint Criterion: The number of people in the proximal area surrounding each facility is never greater than a specified number.

Types of accessibility:

①Topological Accessibility: Illustrates whether two points in space are physically connected by a transport system, thus enabling movement to take place between them.

②Relative Accessibility: Measures the degree of connectivity or accessibility between places.

③Integral Accessibility: Measures the accessibility of a site to a number of other sites or activities.

The basic principle lies in that the impact of one location on another is directly proportional to

Relative Accessibility

j

i $A_i=C_{ij}$

A. Travel time to nearest health clinic, distance to Central Business District

Integral Accessibility

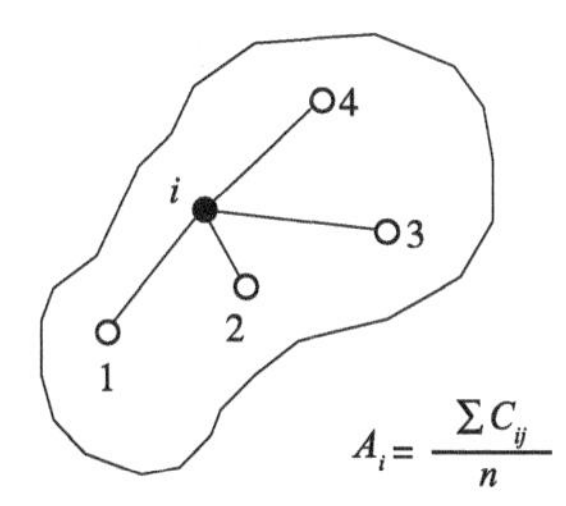

$A_i = \frac{\Sigma C_{ij}}{n}$

B. Mean travel time to all health clinics, in the region mean distance to all other zones

Figure 10-12 Example of Relative and Integral accessibility

its supply (attractiveness) but *inversely proportional* (反比) to its distance. The example is shown as Figure 10-13.

		Destination	
		M_1	M_2
Origin	A	0.5	0.7
	B	1.6	2.1

A. The distance between the origins A and B and destinations M_1 and M_2 in kilometres

Attraction	
M_1	M_2
3.0	5.0

Production	
A	B
2	3

B. To compute accessibility, each market is given an index of how attractive it is and the population (or production) at each location is determined

		Destination	
		M_1	M_2
Origin	A	24.0	20.4
	B	3.5	3.4

C. The raw interations between the two locations A and B and centres M_1 and M_2 result in a 2 by 2 matrix

Figure 10-13 Example of Computing Origin Accessibility and Spatial Interaction

The total accessibility can be expressed as follows:

$$P_i = \sum_{j=1}^{n} \frac{W_j}{d_{ij}^{\beta}} \tag{10-1}$$

Where, P_i is the accessibility at point i; W_j is the attractiveness of location j; d_{ij} is the distance between locations i and j; β is the exponent for distance decay; n is the number of locations in a region.

Assuming a distance decay exponent of 2, the accessibility of A to M_1 is computed as:

$$M_1 = \frac{\text{attractiveness index of market centre}}{(\text{distance between location and market})^2} \tag{10-2}$$

Spatial Interaction refers to "How accessible is one location to all other locations?" and it is computed as:

$$\text{Interaction} = \frac{\text{production} \times \text{attractivencess}}{(\text{distance between origin and destination})^2} \quad (10\text{-}3)$$

Vocabulary

allocation[ˌælə'keɪʃn] *n.* 分配；配置；安置
accessibility[əkˌsesə'bɪləti] *n.* 可以得到；易接近
conduit['kɔndɪt] *n.* 导管；管道；沟渠
critical['krɪtɪkl] *adj.* 批评的；挑剔的；决定性的；危险的，临界的
criterion [kraɪ'tɪəriən] *n.* 标准；准则
decay[dɪ'keɪ] *v.* (使)衰退；(使)腐败；腐烂 *n.* 衰退；腐败；腐烂
impedance [ɪm'piːdns] *n.* 阻抗
linear['lɪniə(r)] *adj.* 直线的；长度的；线性的
non-planar['nɒnpl'ænə] *adj.* 空间的；非平面的；曲线的
optimum['ɒptɪməm] *adj.* 最佳的；最适宜的 *n.* 最适宜
proportional[prə'pɔːʃənl] *adj.* 成比例的；相称的 *n.* 比例项
perception[pə'sepʃn] *n.* 观念；洞察力；认识能力
warehouse['weəhaʊs] *n.* 仓库 *vt.* 存入仓库
threshold['θreʃhəʊld] *n.* 门槛；开端；界限；入口

Questions for Further Study

1. How to identify applications of network analysis?
2. How to explain the performance of networks and network operations?
3. Try your best to demonstrate an operator to find the shortest path.
4. Explain design issues in the construction of networks.

Chapter 11

GIS as A Decision Support Tool

GIS can aid decision-makers to make right and reliable choices such as selecting the best possible site for agricultural development. The specific criteria have been analyzed and prepared by the specialists and the Department of Agriculture. The criteria analyzed and expressed with the mapping technique are used to assess the suitability and seek potential and suitable locations for farming. When the potential sites have been found in several locations, appropriate techniques are applied to make the best choice by GIS. Therefore, GIS can help us to answer and solve some specialized geographical problems.

11.1 Concept

11.1.1 Decision Support System

Every day a lot of decisions need to be made in different sectors around the world. However, better decisions will change an individual's life, and big ones may even have an impact on a country's future.

The relevant concepts of the term Decision Support System (DSS, 决策支持系统) are listed as follows:

①DSS is a conceptual framework for a process of supporting managerial decision-making, usually by modelling problems and employing quantitative models for solution analysis (Turban, 2005).

②DSS is a new type of computer science based on Management Information System and operational research (陈文伟, 2000).

③DSS is a model base (模型库) management system which effectively organizes and stores numerous models. Organic integration of model base and database is built by human-machine interactive function.

④ DSS integrates data processing of Management Information System and numerical calculation function of the model to produce information needed for users.

⑤Aiming to provide a decision, DSS has an "auxiliary" impact on a decision-maker and helps decision-makers to make high-level decisions (王家耀, 2003).

⑥Associated with other concepts, automated Decision Support System (ADSS) shares its idea in the real world such as DSS and Business Intelligence (BI). Besides, ADSS implements different technologies related to AI, and more reliable solutions are sought.

DSS has to face the era of big data and conceives how to meet new demands and solve these problems for DSS in the future by integrating technologies of big data and cloud computing.

11. 1. 2 Spatial Decision Support

Nowadays, spatial data, mathematical models and expert knowledge are more and more *incorporated*(合成一体的) into the decision-making processes. As soon as spatial data is implemented in DSS, GIS functionalities get an important role. These functions enable users to generate spatially diverse decisions. Spatial Decision Support System (SDSS) is designed to help decision-makers to solve complex spatial problems. SDSS provides an opportunity to integrate various analytical models, visualize and evaluate the used models and develop management strategies.

Modern SDSS combines functionalities and modules of GIS, DSS, RS, models and expert knowledge. Furthermore, these system allows the loose coupling of numerical, statistical or knowledge-based expert models, to meet the requirement of being all-inclusive decision support tools.

However, SDSS still hasn't got a clear unified definition. It is generally acknowledged that SDSS is the computer system of processing and analyzing the geographical spatial data and visual modelling to help users make decisions on semi-structured or unstructured spatial problems through the processing and analysis of geospatial data and visual modeling. SDSS is the product of two kinds of technology integrated with DSS and GIS (Wu, 1998) (普遍认为，空间决策支持系统是通过地理空间数据和可视化建模的处理与分析，应用空间数据、应用模型、软件工具，协助用户对半结构或非结构空间问题进行决策的计算机系统。SDSS 是 DSS 和 GIS 两种技术耦合的产物). SDSS consists of spatial decision support, spatial database and several elements of interdependence and interaction (SDSS 由空间决策支持、空间数据库及若干相互依存、相互作用的要素构成).

Spatial decision making is a decision-making process based on geographical (or spatial) relationships and phenomena. It requires a comprehensive analysis of spatial and attributes data.

11. 2 General Approach for Spatial Decision Support

11. 2. 1 Identification of the Problem

Spatial decision support provides solutions to decision-makers because its goal is to solve some objectively practical problems. To avoid different interest groups or individual feelings being involved in and thus influencing problem solving. It is necessary to identify the spatial issues in advance and analyze the problems, such as: What is the object of the problem? What is the purpose of the problem? What is the key factor of influence on the problem solved?

11.2.2 Determination of the Objective

In the real world, decision-making problems are rarelya net single objective. Instead they are inherently multi-objective, *consensus* (共识) rarely exists on the relationships between the various objectives.

If there is more than one objective, then define the relationship between objectives by quantifying them with *commensurate*(适当的) term, i.e. express each objective in the same units. Then the objectives can be simplified into a single objective for analysis. At the same time, decision-makers should agree on the relative importance of the *commensurable* (可公度的，成比例的) objects, and theoretical assumptions are necessary for solving practical problems, e.g. Weber's theory, Geo-statistical Analyst. However, these assumptions do not necessarily held.

When choosing a more applicable and suitable objective indicator system, the principles of the following aspects should be considered:

① *Importance* (重要性): Each selected factor for the aim of assessment has an important influence.

② *Difference* (差异性): Spatial *variability* (变异性) of the selected factor is bigger in the evaluation area.

③ *Accessibility*(可获得性): The data about criteria can be obtained by employing the conventional approach.

④ *Stability* (稳定性): Selected factors possess (具有) relative stability in time series, and the results of the evaluation hold long validity.

11.2.3 Solution of Problem

11.2.3.1 Non-inferior Solution Set Generating Techniques

Non-inferior solution techniques are used when a very large number of options exist.

Generally, single-objective optimization is used to determine the optimal or near-optimal solutions according to the sole objective function (单目标优化根据单一目标函数用以确定最佳或近乎最佳的解决方案). However, there is always more than one solution for problem solving in the real world according to different criteria or constraints. Therefore, multi-objective optimization is to determine the non-inferior (or non-dominated) one among all the solutions (多目标优化的目的是在不同解决方案中确定非劣解). For example, solution A is non-inferior to solution B, only if solution A is no worse than solution B in all the objectives. It also means that solution A dominates solution B or that solution B is a non-dominated solution A. A global non-dominated solution is defined as the solution no worse than any other feasible solution in all the objectives. The task of multi-objective optimization is to search for all the global non-dominated or non-inferior solutions, also known as the *Pareto-optimal set* (帕累托最优解集) or *Pareto-optimal front* (帕累托最优前沿解).

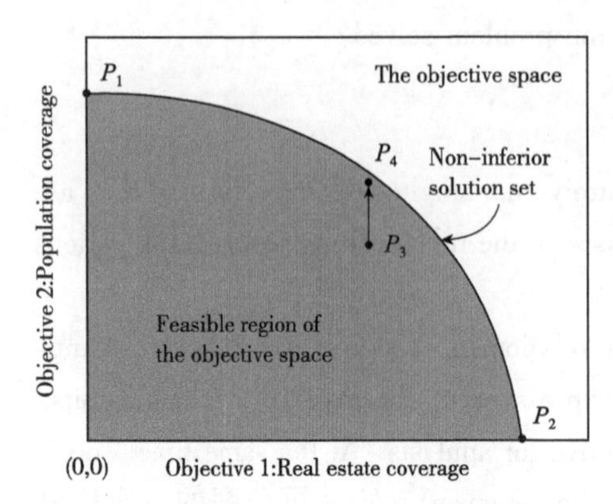

Figure 11-1 Non-inferiority

Figure 11-1 shows the objective space for a two objective problem-the fire station problem.

Two objectives, real estate coverage and population coverage, are represented by the two axes of the graph. The shaded area represents the set of all possible feasible locations (subject to constraints of cost, distance etc). P_1 represents the solution that optimizes coverage of population alone, and P_2 represents the solution that optimizes coverage of real estate.

①A site is non-inferior if an alternative site doesn't exist or the realization of one objective does not take the cost of another one (如果不存在替代站点，且一个目标的实现不会引起另一个目标的损失，则这个站点是非劣效性的).

P_3 represents a feasible solution(可行解) that is not non-inferior. P_3 can move vertically to improve population coverage without changing real estate coverage. Solutions exist which are better than P_3 on one axis (one objective) without necessarily being worse on the other axis.

The curved line represents the set of non-inferior solutions P_4. To improve P_4 for one objective requires a loss on the other.

②The set of non-inferior solutions is the best compromise solution (非劣解集是最好的折衷解决方案) or the *trade-off curve* (取舍选择曲线) in Welfare Economics. Any point on the trade-off curve represents a point of Pareto optimality which indicates a solution point where no objective can be improved without a *sacrifice* of another objective (取舍选择曲线上的任一点均为帕累托最优状态点，在这些点上没有一个目标的实现能在不牺牲其他目标的基础上进行改进).

P_4 cannot move vertically to improve population coverage and it must slide along the trade-off curve. Movement upwards along the curve will imply a change (loss) in the real estate objective. Therefore P_4 is a Pareto optimal or a non-inferior solution point.

Solutions dete rmine the set of all possible sites for the new fire station that represent non-inferior set for each solution, examine the trade-off between covering more lives and more real estate. Decision-makers are frequently interested in the trade-off relationship between the various criteria.

11. 2. 3. 2 Preference Oriented Approaches

Preference oriented approach(偏好导向法) derives a unique solution by specifying goals or preferences. This technique assumes the set of possible solutions is known and small. A typical example is *goal programming*(目标规划).

Goal programming, as one of the oldest and most well-known multi-objective research methods, is generally *utilized* (利用) when there are several competing goals or objectives. Taking land suitability assessment for instance, is used to identify which best suits a set of development or search criteria out of a set of parcels of land. The overall aim is to meet all the criteria or goals to the full extent and to choose the most desirable plan from a set of possible options.

(1) Choose Criteria and Assign Weights

Suppose there are 4 sites to be evaluated, and 8 criteria have been identified. These likely reflect the opinions of different experts, different thoughts, and different objectives. Simultaneously, weights have been given to each criterion to identify its importance (weights must sum to 1). Each site has been ranked on each of the criteria.

(2) Build a Concordance Matrix

Take each ordered pair of alternatives, e. g. sites *A* and *B*, pair *AB*. For each criterion, assign the pair to one of three sets: ①where *A* beats *B* (concordance set), e. g. criteria 2 (wt= 0. 1), 4 (0. 2), 6 (0. 1), 8 (0. 1); ②where *B* beats *A* (discordance set), e. g. criteria 1 (wt=0. 1), 3 (0. 1), 7 (0. 1); ③where *A* and *B* tie (tie set), e. g. criteria 5 (wt=0. 2) .

Sum up the weights of the cases in each set. For example, actual weights for pair *AB* should be: concordance set-0. 5; discordance set-0. 3; tie set-0. 2.

Then determine concordance for each pair by summing up the weights for criteria assigned to concordance set plus the half of sum of weights for criteria in tie set. For pair *AB*: 0. 5 + 0. 1 = 0. 6, it indicates a slight preference for *A* over *B* across all criteria.

Create a matrix of concordance for each pair. The row is the first in pair and the column is the second. Row in total yields an index of preferability. The larger the index is, the more preferred the option is. In conclusion, the overall criterion is that site *D* is preferred to site *C* which is preferred to site *A* and site *A* is preferred to site *B*.

Decision-makers are obliged to specify goals and relative weightings for the different criteria. They use relative weightings to find the most preferred site. Relative weightings can be *alter*ed (改变) to assess the sensitivity of the solution or to reflect different opinions.

11. 2. 3. 3 Weighting Method

Weights can be assigned using the method of *analytic hierarchy process* (AHP, 层次分析法). The AHP is a kind of evaluation method that combines quantitative analysis with qualitative analysis by matrix calculation based on the structure model. Based on the basic principle of AHP, firstly determine relative factors about a complicated environmental problem and their hierarchies structure. Secondly, make certain of their comparative significance by comparing these factors with each other. Finally, determine their weights. Basic AHP procedures are as follows:

Step 1. Develope the weights for the criteria by: developing a single *pair-wise comparison matrix* (成对比较矩阵) for the criteria; multiplying the values in each row together and calculating the *n* th root of said product; normalizing the aforementioned*n* th root of products to get the appropriate weights; calculating and checking the *consistency ratio* (CR, 一致性比率).

Step 2. Develope the rating for each solution alternative on each criterion by: developing a pair-wise comparison matrix for each criterion, with each matrix containing the pair-wise comparisons of the performance of solution alternatives on each criterion; multiplying the values in each row together and calculating the$n$ th root of said product; normalizing the aforementioned nth root of product values to get the corresponding ratings; calculating and checking the consistency ratio (CR).

Step 3. Calculate the weighted average rating for eachsolution alternative. Choose the one with the highest score.

11.3 Application of Spatial Decision Making

A typical case of spatial decision making is farmland suitability assessment which aims at identifying and deciding an ideal parcel of land for agriculture development to maximize economic efficiency and productive benefit.

11.3.1 Problem Identification

To identify which one is the best, the better or barely suitable of a set of parcels of land, the overall aims are to meet all the criteria or goals to the greatest extent, to choose the most desirable plan from a set of possible options for developing farming, to find out basic production capacity, to supply scientific reference and technique supporting for scientific fertilization, improvement of soil quality, adjustment of agricultural structure, sustainable development of agriculture, as well as to protect and construction of farmland quality.

Three sub-goals: soil physical properties, soil chemical properties, soil management are extracted (Table 11-1).

Table 11-1 Discriminant Matrix of Goal Layer (目标层判别矩阵)

Sub-goals	Soil Physical Properties	Soil Chemical Properties	Soil Management
Soil physical properties	1.0000	0.5235	1.0039
Soil chemical properties	1.9102	1.0000	1.9176
Soil management	0.9961	0.5215	1.0000

11.3.2 Data Acquisition

To achieve the goals using the spatial decision-making method, it is necessary to acquire spatial information such as current land usemap, soil type map, DEM (Digital Elevation Model), administrative region map and soil nutrient data.

11.3.3 Criterion Selection and Weight Determination

According to the characteristics of cultivated land resources, and three sub-goals of the suitability of cultivated land, eleven indicators are selected as the criteria for cultivated land suitability assessment, as shown in Table 11-2.

Table 11-2 Criteria of the Cultivated Land Suitability Assessment

Sub-goals	Criteria				
Soil Chemical Properties	Organic Matter	Available Phosphorous	Quick-acting Potassium	Zinc	Salt Content of the Arable Layer
Soil Physical Properties	Soil Texture	The Available Thickness of Soil	Soil Profileconstitution	Topographic Position	
Soil Management	Irrigation Guarantee Rate		Covering Rate of Forest Land		

(1) Discriminant Matrix Calculation for the Criteria

Discriminant matrix of the criteria for soil physical properties, soil chemical properties and soil management is shown as Table 11-3, Table 11-4 and Table 11-5, respectively.

Table 11-3 Discriminant Matrix of the Criteria for Soil Physical Properties

Criteria	Topographic Position	Soil Profileconstitutions	Available Thickness of soil	Soil Texture
Topographic Position	1.0000	1.1765	0.7500	0.9231
Soil profileconstitutions	0.8500	1.0000	0.6375	0.7846
Available thickness of soil	1.3333	1.5686	1.0000	1.2308
Soil texture	1.0833	1.2745	0.8125	1.0000

Table 11-4 Discriminant Matrix of the Criteria for Soil Chemical Properties

Criteria	Organic Matter	Available Phosphorous	Quick-acting Potassium	Zinc	The Salt Content of the Arable Layer
Organic Matter	1.0000	1.5941	3.1569	2.6393	1.4000
Available Phosphorous	0.6273	1.0000	1.9804	1.6557	0.8783
Quick-acting Potassium	0.3168	0.5050	1.0000	0.8361	0.4435
Zinc	0.3789	0.6040	1.1961	1.0000	0.5304
The Salt Content of the Arable Layer	0.7143	1.1386	2.2549	1.8852	1.0000

Table 11-5 Discriminant Matrix of the Criteria for Soil Management

Criteria	Irrigation Guarantee Rate	Covering the Rate of Forest Land
Irrigation Guarantee Rate	1.0000	2.6429
Covering the Rate of Forest Land	0.3784	1.0000

(2) Combined Weight

The combined weights of assessment factors are as follows (Table 11-6).

Table 11-6 Combined Weights of Assessment Factors

	Hierarchy C			Combined Weight
Hierarchy A	Soil Physical Properties B_1	Soil Chemical Properties B_2	Soil Management B_3	$\Sigma B_i A_i$
	0.2560	0.4890	0.2550	—
Topographic Position	0.2344	—	—	0.0600

(续)

	Hierarchy C			Combined Weight
Soil Profileconstitution	0. 1992	—	—	0. 0510
The Available Thickness of Soil	0. 3125	—	—	0. 0800
Soil Texture	0. 2539	—	—	0. 0650
Soil Organic Matter	—	0. 3292	—	0. 1610
Available Phosphorous	—	0. 2065	—	0. 1010
Quick-acting Potassium	—	0. 1043	—	0. 0510
Zinc	—	0. 1247	—	0. 0610
Salt Content of the Arable Layer	—	0. 2352	—	0. 1150
Irrigation Guarantee Rate	—	—	0. 7255	0. 1850
Covering the Rate of Forest Land	—	—	0. 2745	0. 0700

11. 3. 4 Construction of Membership Function

(1) Selection of Membership Function

According to the type of assessment index in the study area, function models are divided into four kinds, i. e. limiting top type, limiting bottom type, linear type and conceptual type. These models are selected to study the relationship between the expression assessment index and the productive potential of cultivated land. The four types are shown as follows:

①Limiting Top Type Function:

$$Y_i=\begin{cases}0 & u_i<u_t \\ & u_t<u_i<c_i(i=1,\ 2,\ \ldots,\ n) \\ 1/[1+a_i(u_i-c_i)^2] & c_i<u_i\end{cases} \tag{11-1}$$

In the equation, Y_i is the score of the ith factor; u_i is an observation of sample; c_i is the standardized index; a_i is coefficient; u_t is lower limiting value.

②Limiting Bottom Type Function:

$$Y_i=\begin{cases}0 & u_t<u_i \\ 1/[1+a_i(u_i-c_t)^2] & c_i<u_i<u_t(i=1,\ 2,\ \ldots,\ m) \\ 1 & u_i<c_i\end{cases} \tag{11-2}$$

In the equation, Y_i is the score of the ith factor; u_i is an observation of sample; c_i is the standardized index; a_i is coefficient; u_t is upper limiting value.

③Linear Type Function:

$$Y_t=ax_i+b \tag{11-3}$$

In the equation, Y_i is the score of the *ith* factor; x_i is an observation of sample; a and b are coefficients.

④Conceptual Type Function: These kinds of index are characterized as qualitative and comprehensive. There is a nonlinear relationship between the index and cultivated land productive potential.

(2) Fitting of Membership Function

According to theassessment value given by experts and the corresponding assessment index, limiting top type, limiting bottom type, linear type and conceptual type models are applied to regression fitting, thus regression function model is constructed. The regression function model should reach a significant level by fitting test.

Among eleven assessment factors, the six factors are quantitative indexes which can be used to simulate the calculation model. The other five are conceptual indexes. Membership degree is proposed by expert's experience according to the correlation of assessment index and cultivated land productive potential (Table 11-7 and Table 11-8).

Table 11-7 Quantitative Assessment Factors From Experts

Factors	Items	Expert Evaluation Value					
Organic Matter(g/kg)	Grade	≥30	20-30	10-20	6-10	<6	—
	Expert Evaluation Value	1	0.9	0.8	0.6	0.4	—
Available Phosphorous (mg/kg)	Grade	≥30	20-30	10-20	5-10	<5	—
	Expert Evaluation Value	1	0.9	0.7	0.5	0.3	—
Quick-acting Potassium (mg/kg)	Grade	≥250	200-250	200-150	150-100	50-100	<50
	Expert Evaluation Value	1	0.8	0.6	0.4	0.3	0.1
Salt Content of Arable Layer (mg/kg)	Grade	<0.25	0.25-0.4	0.4-0.8	0.8-1.2	1.2-1.5	≥1.5
	Expert Evaluation Value	1	0.9	0.8	0.6	0.4	0.2
Available Thickness of Soil (cm)	Grade	100	80-100	60-80	40-60	20-40	<20
	Expert Evaluation Value	1	0.9	0.8	0.6	0.4	0.1
Zinc (mg/kg)	Grade	≥1	0.5-1	0.3-0.5	<0.3	—	—
	Expert Evaluation Value	1	0.8	0.6	0.4	—	—

Table 11-8 Non-quantitative Assessment Factors from Expert

Factors	Items	Expert Evaluation Value				
Topographic Position	Grade	Alluvial-proluvial Fan Central	Alluvial-proluvial Fan Upper Part	Alluvial-proluvial Fan Lower Part	mountain Sloping Field	Desert Edge
	Expert Evaluation Value	1	0.6	0.6	0.5	0.4
Irrigation Conditions	Grade	Better	Good	General	Bad	
	Expert Evaluation Value	1	0.75	0.5	0.3	

(续)

Factors	Items	Expert Evaluation Value				
Soil Texture	Grade	Loam	Sandy Loam Soil	Clay	Sandy Soil	
	Expert Evaluation Value	1	0.8	0.7	0.6	
Soil Profile Constitutions	Grade	Upper Sand and Lower Clay	Loam of All	Sand Covered	Leakage Sand	Sand of All
	Expert Evaluation Value	1	0.8	0.6	0.5	0.4
Covering Rate of Forest Land	Grade	Four Sides Forest	Three Sides Forest	Two Sides Forest	One Side Forest	Nothing
	Expert Evaluation Value	1	0.8	0.6	0.4	0.3

11.3.5 Integrating Criteria and GIS

(1) Spatial Visualization of Point Data(点数据空间可视化)

Spatial positions of field sampling points are recorded bylongitude and latitude using GPS. According to geostatistics principle, the map of spatial visualization can be generated by using Kriging interpolation method(Figure 11-2).

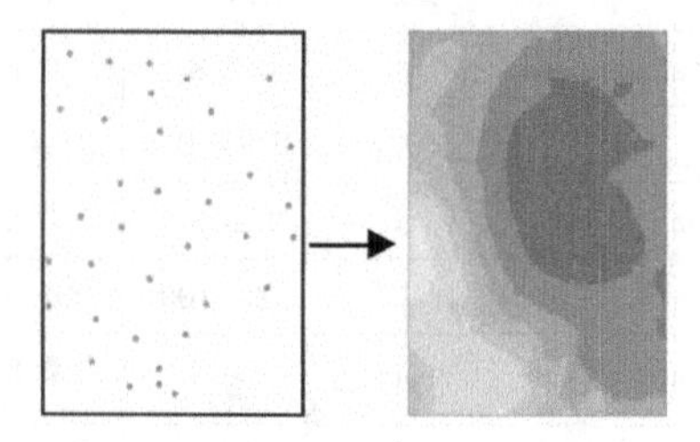

Figure 11-2 Spatial Visualization of Point Data

(2) Calculating Farmland Productivity Composite Index

The following model is used to calculate the composite index of farmland productivity:

$$IFI = \sum F_i \times C_i \quad (i = 1, 2, \ldots, n) \tag{11-4}$$

In the equation, *IFI* means the composite index of farmland productivity; *F* is the score of the ith factor; C_i is the combined weight of the ith factor.

The resulting map of farmland productivity in the study area can be as follows(Figure 11-3):

Therefore, it is more appropriate to identify and maintain the multiple criteria of problems in the real world for analysis and decision making. Decision-makers are frequently interested in the trade-off relationship between the various criteria. This allows them to make the final decisions in a policy environment.

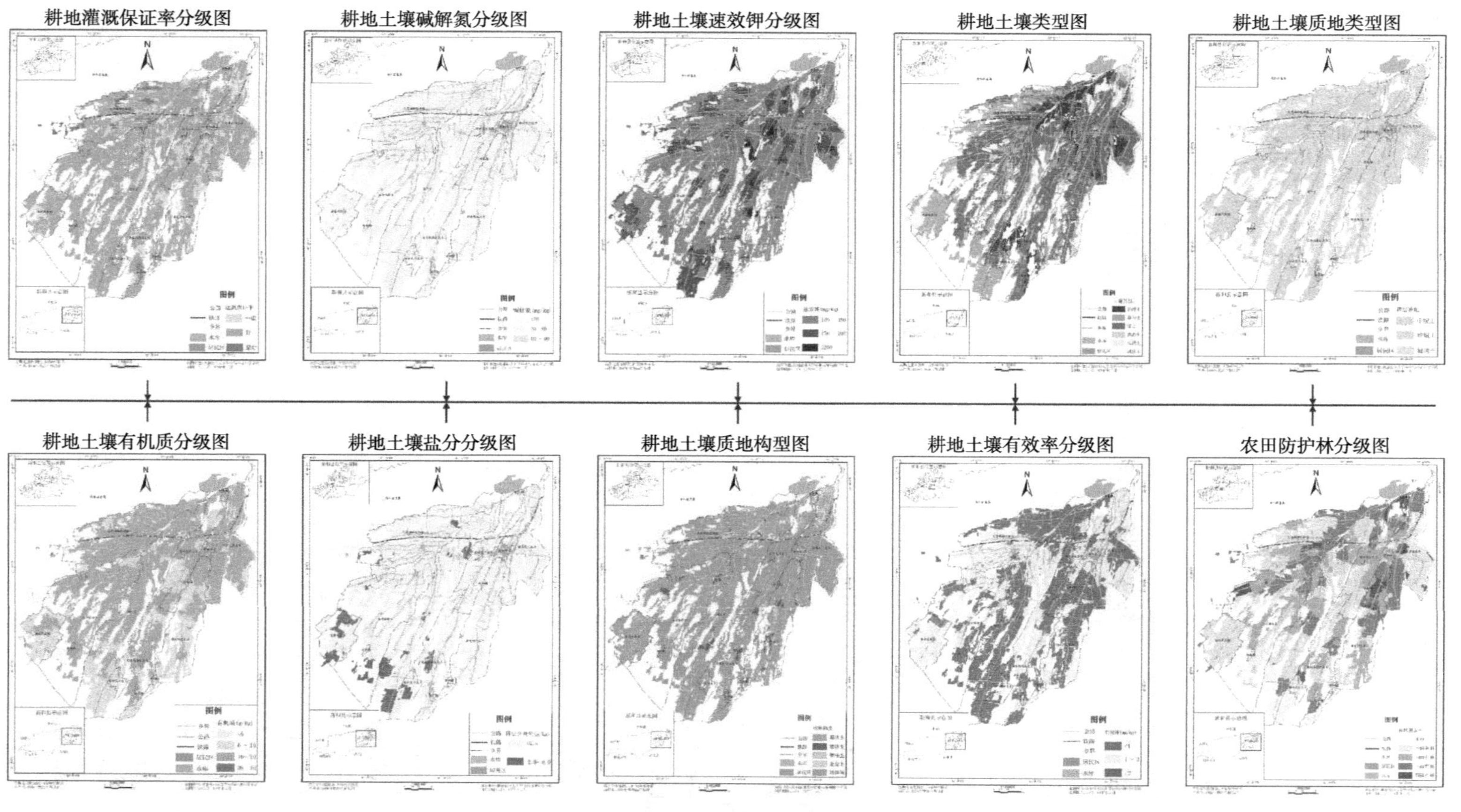

Figure 11-3 A Farmland Suitability Model

Vocabulary

assumption/ə'sʌmpʃən/ *n.* 假定；设想；担任(职责等)；假装
alternative site 替代站点
analytic hierarchy process (AHP) 层次分析法
all-inclusive decision support tools 全方位决策支持工具
big data 大数据
business intelligence 商业智能
constraint /kən'streɪnt/ *n.* 约束；强制；约束条件；对感情的压抑；虚情假意；[计算机]限制
commensurate /kə'menʃərət/ *adj.* 相称的；成比例的；适当的
criterion /kraɪ'tɪəriən/ *n.* 标准，准则
cloud computing 云计算
conceptual type 概念模型
consistency ratio 一致性比率
concordance matrix 一致性矩阵
cultivated land 耕地
decision support tool 决策支持工具
decision support system 决策支持系统
decision-maker 决策者
decision-making 决策
discriminant matrix 判别矩阵
model base 模型库
knowledge-based expert-models 基于专家知识模型
farmland productivity composite index 耕地生产力综合指数
fitting /'fɪt. ɪŋ/ 拟合
goal programming 目标规划
Kriging interpolation method 克里金插值方法
limiting top type 界上型
limiting bottom type 界下型
linear type 线性类型
multi-objective 多目标
non-inferior 非劣解
Pareto optimality 帕累托最优
quantitative models 定量模型
quantitative assessment 定量评价
soil physical properties 土壤物理属性
soil chemical properties 土壤化学属性

soil management　土壤管理
spatial visualization　空间可视化
spatial decision support　空间决策支持
spatial Decision Support System(SDSS)　空间决策支持系统
spatial variability　空间变异性
weighting method　加权法

Questions for Further Study

1. What is spatial decision making?
2. What is basic AHP procedure?
3. What is the weight method?
4. How to integrate criteria with GIS?
5. Try your best to solve a problem around yourself by leaning knowledge of this chapter.

Chapter 12

Main GIS Application Area

Nowadays, GIS technologies have been widely applied to diverse fields to assist humans in analyzing various types of geospatial data and dealing with complex situations. In business, education, natural resources, tourism and transportation and so on, GIS plays an essential role in helping people to collect, analyze the related spatial data and display data in different formats (当前，GIS 技术已经被广泛应用到许多不同的领域用以帮助人们分析不同类型的地理空间数据和处理复杂的状况。无论在商业、教育、自然资源、旅游还是交通等领域，GIS 均发挥着帮助人们收集、分析相关的空间数据并用不同形式将数据表达出来的基础性作用).

12.1 Basics of GIS Applications

GIS technology, data structures and *analytic* (分析的) techniques are gradually being incorporated into a wide range of management and decision-making operations. Numerous case studies of the GIS application are hot topics in many journals and conferences of the natural and social sciences (在自然科学和社会科学领域的诸多期刊、会议的热门话题中，我们都能找到 GIS 应用的大量实例). To understand the range of applicability of GIS, it is necessary to characterize the multitude of applications in some logical way so that *similarities and differences* (相似性和差异性，异同) between approaches and needs can be examined. An understanding of this range of needs is critical for those who will deal with the procurement and management of GIS.

12.1.1 Functional Classification

One way to classify GIS applications is to identify the functional characteristics of the systems. Considerations should be taken as follows: ① characteristics of the data, such as themes, precision, data model, etc.; ② possible GIS functions that may be adopted in the application, such as address matching, overlay, and so on; ③ potential GIS products the application would support, such as one-time video maps, hardcopy maps and so on.

It is worth noting that this kind of classification based on these considerations is quickly becoming fuzzy with GIS as a flexible tool that has great capacities to integrate data themes, functionality and output.

12.1.2 Decision Support Tool

Another way to classify GIS is the kinds of decisions supported by the GIS, such as simple decisions as necessary in the process of making decisions; and complex decisions typically involving issues like these:

①Uncertainty-many facts may not be known.

②Complexity-many interrelated factors need to be considered.

③High-risk consequences-the impact of the decision may be significant.

④Alternatives-each has its own set of uncertainties and consequences.

⑤Interpersonal issues-it can be difficult to predict how other people will react.

Decision support is an excellent goal for GIS, however decisions range from major ones to minor ones. It is sometimes difficult to know when GIS should be used to make decisions except in cases of major decisions.

12.1.3 Groups of GIS Activity

GIS is used by nearly all people who deal with spatial information, including the public, administrative personnel, academics and professionals. Each group has a distinctive educational and cultural background with varied *interests* (嗜好) and *priorities* (特权). As a result, each identifies itself with particular ways of approaching particular sets of problems, though sometimes interactions occur between them.

The core groups of GIS activity can be comprised of: ①mature technologies which interact with GIS, sharing their technology and data with the different groups, such as surveyors, engineers, cartographers, remote sensing scientists, etc.; ②management and decision-making groups, such as resources managers, *urban planners* (城市规划师), *municipal officials*(市政官员), facilities managers, etc.; ③science and research activities at universities and government labs, these groups of GIS activity are seeking to find distinctions and similarities between them.

12.2 Scopes of GIS Applications

GIS technology has *emerged* (出现) just for a few short years, but it has been successfully applied to municipal engineering, enterprise decision-making, resource management, transportation, health care, post & telecommunications, public security & first aid, marketing, finance & insurance, petroleum chemical industry, water conservancy & electric power, environmental protection, tourism, education and scientific research (GIS技术出台虽短短几年，但已成功应用在市政工程、企业决策、资源管理、交通运输、医疗保健、邮电通信、公安急救、市场销售、金融保险、石油化工、水利电力、环保旅游、教育科研等各个方面). The main application areas of GIS are introduced briefly below.

12. 2. 1 GIS in Cartography

There are two areas: ①automation of the map-making process; ②production of a new form of maps resulting from the analysis, manipulation of data. The second is closer to the concept of GIS although both of them use similar technology.

Cartography(地图制作) dates back to thousands of years ago when no paper came out, but the main visual display principles were developed during the paper era and thus many digital cartographers today still use the *terminology* (术语), conventions, and techniques from the paper era.

Conventionally, cartography (or map) is used as chart for *terrestrial areas* (陆地) and *marine areas* (海域). In cartography, artistic tools and techniques such as colouring, symbolizing, lettering, etc. are employed. These visual devices, tools, and methods form iconography, which is a concise expression of the important culture (地图中的这种艺术简明地表达了一种重要的文化). Cartography *concerns* (涉及) the art, science, and techniques of making maps or charts. In a sense, cartographers are the artists.

12. 2. 2 GIS in Surveying and Engineering

There are many advantages of GIS in surveying, engineering, construction and maintenance. With decreasing types of software used, training and *licensing* (授权) costs will be reduced, *compatibility issues* (兼容的问题) will be minimized, and the chance of error will be reduced when data are translated from one system to another. *Project personnel* (项目成员) can more easily *cross-train* (跨界) to other work functions, The operational GIS system stores the information collected from concept through construction, giving system operators much more data to operate and manage their system efficiently.

The relationship betweensurveyors and civil engineers is of critical importance to the GIS market. *Civil engineers* (土木工程师) design and build this infrastructure while surveyors collect and deliver highly accurate geospatial data, fueling multiple industries and levels of government. Using GIS technology, surveyors and civil engineers play a key role in the GIS data for improvement process and contribute their *expertise* (专长) in design, measurement, accuracy, and precision to improve the quality/accuracy of GIS data.

12. 2. 3 GIS in Remote Sensing

Remote sensing is the art and science of making measurements of the earth using sensors on airplanes or satellites. These sensors collect data in the form of images and provide specialized capabilities for manipulating, analyzing, and visualizing those images. Remote sensed imagery is integrated within GIS.

Remote sensing technology in recent years has proved to be of great importance in acquiring a variety of data concerning the identification of *plant community* (植物群落), land cover, *biomass* (生物量), and qualitative estimation of land use change, *suspended sediment concentration* (悬浮

沙量浓度), crop production, surface temperature, for effective resources management.

12.2.4 GIS in Resource Management

GIS can be used to indicate where a problem may occur or what impacts may be resulted due to proposed actions or resource exploitation, allocation and/or development. GIS is helpful to deal with the recognition of problems within specific socio-economic and environmental settings.

12.2.5 GIS in Urban Development

GIS provides planners, surveyors, and engineers with the tools. GIS enables data from a wide variety of sources and data formats to be integrated with a common scheme of geographical referencing, thus providing *up-to-date* (最新) information. In essence, GIS in the plan-making process intends to achieve better planning through better information that necessarily flows from an information system. No matter how large or small a community is, planners must deal with spatial information: parcels, zoning and land use data, addresses, transportation networks, and housing stock. As a planner, he/she also study and keep track of multiple urban and regional indicators,forecast future community needs, and plans accordingly to *guarantee the quality of life* (保证生活质量) for everyone in *livable* (宜居的) communities.

12.2.6 GIS in Facilities Management

GIS integrates with the top *facilities management* (FM, 设施管理) software and *consulting firms* (咨询公司), makes it easier than ever to extend the life of our FM data. GIS can be used throughout the life cycle of a facility — from deciding where to build to spatial planning.

GIS helps us to: ①make *tramline asset* (有轨交通线路资产) information collection, dissemination, maintenance, and use; ②facilitate better planning and analysis; ③allow efficient sharing of information in and out of the field and provide a comprehensive view of operations.

12.2.7 GIS in Science and Research

Scientific knowledge is created by employing scientific methods on the twin principles of events observation and empirically testable theories (科学知识是通过事件观测和实证检验相结合原则上采用科学方法创造的的). It is common to subdivide science into natural sciences such as *biology*, *environmental science*, and social sciences [人们常将科学分为自然科学(包括生物学、环境学)和社会科学]. Geography is the science that describes and explains the formation process, function and patterns of the earth's surface (地理是用以描述和揭示地表形成过程、功能和格局的科学). GIS is essential to modern geographic information science. Without the help of GIS, it would not be possible to collect large volumes of information about observable events, build and test theories about geographic patterns and processes. Without information system technologies, many interesting *geo-scientific* (地理科学的) problems are *intractable* (难以解决的).

Geographic information science is concerned with the fundamental principles that underlie

GIS such as the basic models, methods, and generally held *tenets* (原理) of geography and geographic information. A basic understanding of geographic information science is essential for all geographic information work. It is a necessary foundation for all GIS data management, analysis, and visualization activities. From the view of emphasizing the acquisition and analysis of scientific computing results, GIS can be taken as a secondary means of scientific research(GIS 可视为科学研究的辅助手段). For such a purpose, it is necessary to use the functionality provided by the GIS common software anda variety of *professional analytical models* (专业分析模块).

12.2.8 GIS in Commercial Service

GIS tools have been used by individuals or organizations in various commercial applications, such as retail, real estate and restaurants. GIS toolset makes it easy to perform the following common tasks: ①create customer-based or store-based trade areas (创建基于客户或基于店面的商业区); ②conduct customer or store surveys (开展客户或商店调查); ③find the best store location (搜寻到最佳店面位置); ④carry out rush-hour transportation analysis (开展交通高峰期分析); ⑤create gravity models to predict potential sales of new stores (创建重力模型来预测新店潜在的销售量).

GIS is in a unique position to unite disparate business disciplines, and it makes impacts on business processes and intra-organizational mechanisms through its potential functions (GIS 具有特殊的地位，可以将不同的商业范畴统一在一起，并通过其潜在功能来影响商业过程和内部组织机制). The issues are included as follows: *facility management* (设施管理), *employee* (*and family*) *management* (雇员管理), *incident mapping* (突发事件制图), *weather mapping* (气象制图), *office relocation* (办公室搬迁), *evacuation* (配送), *threat assessment* (危险评估), *supply chain assessment* (供应链评估), etc.

Analysts (分析员) use GIS in retail applications to answer critical business questions such as: where are my *best customers* (最佳客户)? where can I find more customers? where are my *competitors* (竞争对手)? where should I locate a new store? who is the typical population in favor of my product? These questions can be answered in many ways with basic techniques such as geocoding existing customers and stores and looking for the location. A GIS enables analysts to calculate the distance to stores from each customer and create the *scatter plot* (散点图) to represent this information.

Customer prospecting tools are available within the Business Analyst toolset and have been used for quite some time with GIS. The customer *prospecting wizard* (查找向导) enables users to search for the areas that contain a certain type of customers at various levels of geography. The users can query the data to find the most profitable customers and begin to look for the *demographic characteristics* (人口特征) that differentiate these customers from the average customers such as *income* (收入), *homeownership ratio* (拥有住房比), and so on. The output from the customer prospecting tool is a new layer showing the areas that match the input criteria.

The *hot spot analysis tool* (热点分析工具) is used to find the location of spatial clusters of high and low attribute values. It shows the areas where higher than average values tend to be found near each other and where lower than average values tend to be found near each other. For the customer prospecting output data, this will give us potential locations for new stores. This tool can be run in two modes: the statistics from the analysis report or a new layer file that can be added to a map. This output graphically shows the locations of the hot and *cold spots* (冷点).

12.2.9 GIS in Public Service

GIS can centrally host applications and data, streamline workflow and manage assets, fleet- and other many activities, and enhance customer service (GIS 可以集中托管应用程序和数据，简化工作流程，管理资产、车队及其他很多活动，同时加强客户服务). It has set a solid foundation for supporting the growth of a public service department.

GIS has far-reaching implications in our society. This technology has been explicitly employed in the development and implementation of public policy, in the redistribution of funds, and in identifying communities for target-based intervention. For instance, GIS provides us with information, announcement, public consultation, and public issues management such as emergency service, crowd control, anti-crime actions, public health, and so on.

With GIS, vast amounts of spatial information can be securely stored and managed; data conversion can be done between multiple data sources; data integrity, consistency, and credibility can be ensured; *real-time* (实时) tracking of features and events can be integrated. GIS data can be tied to non-GIS applications. People without GIS knowledge can be allowed to take advantage of geographic data.

12.3 Concluding Comments

GIS is consistently oriented towardhigh performance, low operating cost, more openness and flexibility. This chapter has presented a selection of traditional GIS application fields. Along with the maturity of object-oriented theory, perfection of virtual reality technique, the popularity of network and artificial intelligence systems, WebGIS system integration strategy based on the Internet and *Intranet* (局域网) will be the mainstream of GIS technology in the 21st century. With the quick development of the Internet, it is possible to share GIS data. Establishing user-oriented, resource-sharing and open WebGIS will become the trend of the times. Each GIS user can exchange information with the servers via the Internet and even share updated data online to the globe.

Vocabulary

analytic [ˌænəˈlitik] *adj.* 分析的；分解的
approach [əˈprəutʃ] *v.* 靠近；接近；接洽；要求；达到；动手处理
biomass [ˈbaiəumæs] *n.* (单位面积或体积内的) [生] 生物量
cartography [kɑːˈtɔgrəfi] *n.* 地图制作，制图；制图学，绘图法

compatibility [kəm'pætə'biləti] *n.* 兼容性
consideration [kən'sidə'reiʃən] *n.* 考虑；原因；关心；报酬
diverse[dai'vɔːs] *adj.* 不同的；多种多样的
emerge *vi.* 浮现；摆脱；出现
explicitly [ik'splisitli] *adv.* 明确地；明白地
facilitate [fə'siliteit] *vt.* 促进；帮助；使容易
flexible ['fleksibl] *adj.* 灵活的；柔韧的；易弯曲的
format *n.* 格式；版式；开本
frequent ['friːkwənt] *adj.* 频繁的；经常的
functionality [ˌfaŋkʃə'næliti] *n.* 功能；泛函性，函数性
iconography [ˌaikə'nagrəfi] *n.* 肖像研究；肖像学；图解
infrastructure ['infrəˌstraktʃə] *n.* 基础设施；公共建设；下部构造
integrate ['intigreit] *vt.* 使…完整；使…成整体 *vi.* 求积分；取消隔离；成为一体
integrity [in'tegrəti] *n.* 完整；正直
intervention [ˌintə'venʃn] *n.* 介入；调停；妨碍；干扰
launch [luːntʃ] *vt.* 发射(导弹、火箭等)；发起，发动；使…下水 *vi.* 开始；下水；起飞；体现
multitude ['maltitjuːd] *n.* 群众；多数
mainstream ['meinstriːm] *n.* 主流
manipulation [mə'nipju'leiʃən] *n.* 操纵；操作；处理；篡改
maturity [mə'tjuəriti] *n.* 成熟；到期；完备
procurement [prəu'kjuəmənt] *n.* 采购；获得，取得
recognition [ˌrekəg'niʃən] *n.* 识别；承认，认出；重视；赞誉；公认
redistribution [riːdistri'bjuːʃən] *n.* 重新分配
significant [sig'nifikənt] *adj.* 重大的；有效的；有意义的；值得注意的；意味深长的
terrestrial [ti'restriəl] *adj.* 地球的；陆地的，陆生的；人间的 *n.* 陆地生物；地球上的人

Questions for Further Study

1. Choose an application domain of GIS (e. g. agriculture, land resource inventory) and give specific examples.

2. Report the evolution and the future trends of GIS.

Bibliography

陈文伟，廖建文．决策支持系统及其开发[M]．北京：清华大学出版社，2000.
方裕，周成虎，景贵飞，等．第四代 FGH 软件研究[J]．中国图象图形学报，2001，6(9)：817-823.
龚建雅．地理信息系统与科学基础[M]．北京：科学出版社，2001.
胡鹏，黄杏元，华一新．地理信息系统教程[M]．武汉：武汉大学出版社，2002.
李志林．数字高程模型[M]．2 版．武汉：武汉大学出版社，2000.
梁红莲，刘登忠．GIS 应用现状及发展趋势探讨[J]．物探化探计算技术，2001，23(1)：68-73.
潘瑜春，钟耳顺，赵春江．GIS 空间数据库的更新技术[J]．地球信息科学，2004，6(1)：36-40.
汤国安，刘学军，闾国年，等．地理信息系统教程[M]．北京：高等教育出版社，2007.
汤国安，刘学军，闾国年．数字高程模型及地学分析的原理与方法[M]．北京：科学出版社，2005.
汤国安，赵牡丹，杨昕，等．地理信息系统[M]．2 版．北京：科学出版社，2010.
王家耀，周海燕，成毅．关于地理信息系统与决策支持系统的探讨[J]．测绘科学，2003，28(1)：1-4.
温淑瑶，马占青，周之豪，等．层次分析法在区域湖泊水资源可持续发展评价中的应用[J]．长江流域资源与环境，2000，9(2)：196-201.
邬伦，刘瑜，张晶，等．地理信息系统原理、方法和应用[M]．北京：科学出版社，2001.
吴立新，史文中．地理信息系统原理与算法[M]．北京：科学出版社，2003.
张超．地理信息系统实习教程[M]．北京：高等教育出版社，2000.
张宏，温永宁，刘爱利，等．地理信息系统算法基础[M]．北京：科学出版社，2006.
郑春燕，邱国锋，张正栋，等．地理信息系统原理、应用与工程[M]．2 版．武汉：武汉大学出版社，2011.
ABLER R F. Awards, rewards and excellence: keeping geography alive and well[J]. Professional Geographer, 1988, 40(2): 135-140.
ALLEN P M. Cities and regions as evolutionary complex systems[J]. Geographical Systems, 1997, (4): 103-130.
ARMSTRONG M P, DENSHAM P J. Database organization strategy for spatial decision support systems[J]. International Journal of Geographical Information Systems, 1990, 4(1): 3-21.
ARONOFF S. Geographic Information Systems: a management perspective[M]. Ottawa: WDL Publications, 1989.
BAILLARD C, SCHMID C, ZISSERMAN A, et al. Automatic line matching and 3D reconstructionof buildings from multiple views[C]// ISPRS Conference on Automatic Extraction of GIS Objects from Digital Imagery, 1999, 32: 69-80.
BATTY M, XIE Y, SUN Z. Modeling urban dynamics through gis-based cellular automata[J]. Computers, Environment and Urban Systems, 1999, 23(3): 205-233.

BISHOP I D. Comparing regression and neural net based approaches to modeling of scenic beauty[J]. Landscape and Urban Planning, 1996, 34(2): 125-134.

BLUNDEN W R, Black J A. The land-use/transportation system[M]. 2nd ed. Oxford: Pergamon Press, 1984.

BOOCH G. Object-oriented analysis and design with applications[M]. 2nd ed. Redwood City: Benjamin/Cummings, 1994

BOVY P H L, JANSEN G R M. Network aggregation effects upon equilibrium assignment outcomes: an empirical investigation[J]. Transportation Science, 1983, 17(3): 240-262.

BURROUGH P A. Principles of geographical information systems for land resource assessment[M]. Oxford: Claredon Press, 1986.

CAVER S J. Integrating multi-criteria evaluation with geographical information systems[J]. International Journal of Geographical Information Systems, 1991, 5(3): 321-339.

CHANG K T. Introduction to Geographic Information Systems[M]. 9nd ed. New York: McGraw-Hill Higher Education, 2018.

CHEN Z T, GUEVARA J A. Systematic selection of very important points (VIP) from digital terrain models for constructing triangular irregular networks[J]. Proceedings of Auto-Carto, 1987, 8: 50-56.

CHO H, HONG S, KIM S. Application of a terrestrial Lidar system for elevation mapping in Terra Nova Bay, Antarctica[J]. Sensors, 2015, 15(9): 23514-23535.

CHORLEY R. Some reflections on the handling of geographical information[J]. International Journal of Geographical Information Systems, 1988, 2(1): 3-9.

CHRISMAN N. Exploring geographic information systems[M]. 2nd ed. New York: John Wiley & Sons, 2002.

CHRISTENSEN A H J. Fitting a triangulation to contour lines[G]. Baltimore: Proceedings 8th International Symposium of Computer Assisted Cartography, 1987.

CHUNG M, COBB M, SHAW K, et al. An object oriented approach for handling topology in VPF products[J]. GIS LIS-International Conference-American Society for Photogrammetry and Remote Sensing, 1995, 1: 163-174.

CLARKE K C. Getting started with geographic information systems[M]. 5th ed. Upper Saddle River: Prentice Hall, 2010.

CLARKE M. Geographical information systems and model based analysis: towards effective decision support systems[M]// SCHOLTEN H J, STILLWELL J C H. Geographical information systems for urban and regional planning. Dordrecht: Kluwer Academic Publishers, 1990.

DATE C J. An introduction to database systems[M]. 8th ed. Upper Saddle River: Addison-Wesley, 2004.

DE BY R A. Principles of Geographic Information Systems[M]. 2nd ed. Enschede: The International Institute for Aerospace Survey and Earth Sciences (ITC), 2001.

DEMERS M N. Fundamentals of geographic information systems[M]. 4th ed. New York: John Wiley & Sons, 2008.

DEMERS M N. Fundamentals of geographic information systems[M]. New York: John Wiely & Sons, 1997.

DEMERS M N. GIS modelling in raster[M]. New York: John Wiely & Sons, 2002.

DODGSON J S, SPACKMAN M, PEARMAN A, et al. Multi-criteria analysis: a manual[M]. London: Department for Communities and Local Government, 2009.

EBNER H, REINHARDT W, HÖßLER R. Generation, management and utilization of high fidelity digital terrain models[J]. International Archives of Photogrammetry and Remote Sensing, 1988, 27(B11): 556-565.

EI-NAJDAWI M K, STYLIANOU A C. Expert support systems: an integration of decision support systems, expert systems, and other AI technologies[J]. Communications of the ACM, 1993, 36(12): 55-65.

EVANS I S. An integrated system of terrain analysis and slope mapping[J]. Zeitschrift für Geomorphologie, 1980, 36(Suppl.): 274-295.

EYTON J R. Raster contouring[J]. Geo-Processing, 1984, (2): 221-242.

FABIO R. From point cloud to surface: the modeling and visualization problem[J]. International Archives of Photogrammetry, Remote Sensing and Spatial Information Sciences, 2003, 34(5): W10.

FAZAL S. GIS Basics[M]. New Delhi: New Age International Publishers, 2008.

FOLEY J D, VAN DAM A, FEINER S K, et al. Computer graphics: principles and practice in C[M]. 2nd ed. Upper Saddle River: Addison-Wesley, 1995.

FRANK A U, EGENHOFER M J, KUHN W. A perspective on GIS technology in the nineties[J]. Photogrammetric Engineering & Remote Sensing, 1991, 57(11): 1431-1436.

GAILE G L, WILLMOTT C J. Geography in America[M]. Columbus: Merrill, 1989.

GAILE G L, WILLMOTT C J. Spatial statistics and models[M]. Dordrecht: Springer, 1984.

GOLD C, MOSTAFAVI M A. Towards the global GIS[J]. ISPRS Journal of Photogrammetry & Remote Sensing, 2000, 55(3): 150-163.

GONZALEZ R C, WINTZ P A. Digital image processing[M]. 2nd ed. Upper Saddle River: Addison-Wesley, 1987.

GOODCHILD M F, PARKS B O, STEYAERT L T. Environmental modeling with GIS[M]. New York: Oxford University Press, 1993.

GOODCHILD M, DENSHAM P J. Research initiative 6: spatial decision support systems[J]. Netherlands Heart Journal Monthly Journal of the Netherlands Society of Cardiology & the Netherlands Heart Foundation, 1994.

GUIBAS L, STOLFI J. Primitives for the manipulation of general subdivisions and the computation of Voronoi diagrams[J]. ACM Transactions on Graphics, 1985, 4(2): 74-123.

HANNAH M J. Error detection and correction in digital terrain models[J]. Photogrammetric Engineering and Remote Sensing, 1981, 47 (1): 63-69.

HOBERMAN S. Data modeling made simple: a practical guide for business and professionals[M]. 2nd ed. New Jersey: Technics Publications, 2009.

HOBERMAN S. Data modeling made simple[M]. 2nd ed. Westfield: Technics Publications LLC, 2009.

HUTCHINSON M F. A new procedure for gridding elevation and stream line data with automatic removal of spurious pits[J]. Journal of Hydrology, 1989, 106(3-4): 211-232.

JANKOWSKI P, RICHARD L. Integration of GIS-based suitability analysis and multicriteria evaluation in a spatial decision support system for route selection[J]. Environment and Planning, 1994, 21(3): 323-340.

JANKOWSKI P. Integrating geographical information systems and multiple criteria decision-making methods[J]. International Journal of Geographical Information Systems, 1995, 9(3): 251-273.

JANSSEN R, RIETVELD P. Multi-criteria analysis and geographical information systems: an application to agricultural land use in the Netherlands[M]// SCHOLTEN H J, STILLWELL J. Geographical information systems for urban and regional planning. Dordrecht: Kluwer Academic Publishers, 1990.

JIANG B, YAO X B. Geospatial analysis and modelling of urban structure and dynamics: an overview[J]. Geospatial Analysis and Modelling of Urban Structure and Dynamics, 2010, 99: 3-11.

JIANG B, YAO X B. Geospatial analysis and modelling of urban structure and dynamics[M]. New York: Springer, 2010.

KEENAN P. Using a GIS as a DSS generator[M]// DARZENTAS J, DARZENTAS J S, SPYROU T. Perspectives on DSS. Mytiline: University of the Aegean Press, 1996.

KIM Y H, RANA S, WISE S. Exploring multiple viewshed analysis using terrain features and optimisation techniques[J]. Computers and Geosciences, 2004, 30(9): 1019-1032.

KRAAK M J, ORMERLING F J. Cartography: visualization of geospatial data[M]. 2nd ed. Upper Saddle River: Prentice Hall, 2003.

KWAN A K H, MORA C F, CHAN H C. Particle shape analysis of coarse aggregate using digital image processing[J]. Cement and Concrete Research, 1999, 29(9): 1403-1410.

KÖSTLI A, SIGLE M. The random access data structure of the DTM program SCOP[J]. International Archives of Photogrammetry and Remote Sensing, 1986, 26: 42-45.

LAM N S. Spatial interpolation methods: a review[J]. The American Cartographer, 1983, 10(2): 129-149.

LANG L. GIS for health organizations[M]. New York : ESRI Press, 1999.

LAUDIEN R, BROCKS S, BARETH G. Designing and developing spatial decision support systems by using arcgis engine and Java[C]// 4th GIS Conference & ISPRS Workshop on Geoinformation and Decision Support Systems, 2008.

LEUNG Y. Intelligent spatial decision support systems[M]. Berlin and New York: Springer-Verlag, 1997.

LI Y R, SHI W H, AYDIN A, et al. Loess genesis and worldwide distribution[J]. Earth-Science Reviews, 2020, (201): 102947.

LI Z L, ZHU Q, GOLD C. Digital terrain modeling: principles and methodology[M]. New York: CRC Press, 2005.

LI Z L. Sampling strategy and accuracy assessment for digital terrain modelling [D]. Glasgow: University of Glasgow, 1990.

LONGLEY P A, GOODCHILD M F, MAGUIRE D J, et al. Geographic information system and science[M]. 2nd ed. New York: John Wiley & Sons. 2005.

LONGLEY P A, GOODCHILD M, MAGUIRE D J, et al. Geographic information systems and science[M]. 3rd ed. New York: John Wiley & Sons, 2010.

LONGLEY P A. 地理信息系统与科学[M]. 张晶, 刘瑜, 张洁, 等译. 北京: 机械工业出版社, 2007.

MAGUIRE D J, GOODCHILD M F, Rhind D W. Geographical information systems: [M]. New York: John Wiley & Sons, 1991.

MALCZEWSKI J. GIS and multicriteria decision analysis[M]. New York: John Wiley & Sons, 1999.

MCCALEB M R. A conceptual data model of datum systems[J]. Journal of Research of the National Institute of Standards & Technology, 1999, 104(4): 349-400.

MCCLOY K R. Resource management information systems, remote sensing, GIS and modeling[M]. 2nd ed. Boca Raton: Taylor and Francis, 2005.

MEADEN G J, DO CHI T. Geographical information systems applications to marine fisheries[M]. Rome: Food and Agriculture Organization of the United Nations, 1996.

MOORE J H, WEATHERFORD L R. Decision modeling with Microsoft Excel[M]. 6th ed. Upper Saddle River: Prentice-Hall. 2001.

NEWMAN W M, SPROULL R F. Principles of interactive computer graphics[M]. 2nd ed. New York: McGraw-Hill, 1978.

NG M W. Synergistic sensor location for link flow inference without path enumeration: a node-based approach[J]. Transportation Research(B): Methodological, 2012, 46(6): 781-788.

O'CALLAGHAN J F, MARK D M. The extraction of drainage networks from digital elevation data[J]. Computer Vision, Graphics, and Image Processing, 1984, 28(3): 323-344.

OLIVER G, RUDOLF M. From GISystems to GIServices: spatial computing on the internet marketplace[M]// GOODCHILD M, EGENHOFER M, FEGEAS R, et al. Interoperating geographic information systems. New York: Springer-Kluwer Academic Publishers, 1999.

PEUCKER T K, FOWLER R J, LITTLE J J, et al. The triangulated irregular network[C]//American Society of Photogrammetry. Digital Terrain Models Symposium, 1978.

PEUCKER T K. Computer cartography: commission on college geography resource paper 17[G]. Washington: Association of American Geographers, 1972.

PEUQUET D J, MARBLE D F. Introductory readings in geographic information systems[M]. New York: CRC Press, 1990.

POOYANDEH M, MESGARI S, ALINOHAMMADI A, et al. A comparison between complexity and temporal GIS models for spatio-temporal urban applications[C]// Lecture Notes in Computer Science; 4706. Faculty of Geodesy and Geomatics Eng. K. N. Toosi University of Technology. Computational Science and Its Applications - ICCSA 2007 pt. 2, 2007.

PREPARATA F P, SHAMOS M I. Computational geometry: an introduction[M]. New York: Springer, 1985.

RAMACHANDRAN S, SUNDRAMOORTHY S, KRISHNAMOORTHY R, et al. Application of remote sensing and GIS to coastal wetland ecology of Tamil Nadu and Andaman & Nicobar Group of islands with special reference to Mangroves[J]. Current Science, 1998, 75(3): 236-244.

RIVOLLIER S, DEBAYLE J, PINOLI J C. Shape diagrams for 2d compact sets-part Ⅱ: analytic simply connected sets[J]. The Australian Journal of Mathematical Analysis and Applications, 2010, 7(2): 1-18.

ROSIN P L. Measuring Shape: ellipticity, rectangularity, and triangularity [J]. Machine Vision and Applications, 2003, 14(3): 172-184.

SAATY T L, VARGAS L G. Models, methods, concepts & applications of the analytic hierarchy process[M]. 2nd ed. New York: Springer, 2012.

SAATY T L, VARGAS L G. Models, methods, concepts & applications of the analytic hierarchy process[M]. Dordrecht: Kluwer Academic Publishers, 2000.

SHEN L. GIS-based multi-criteria analysis for aquaculture site selection [D]. Gävle: University of Gävle, 2010.

SIEBER R. A selected bibliography on spatial data handling: data structures, generalization and three - dimensional mapping[J]. EPFL, 1986, 6: 20-35.

SINGH A. Towards decision support models for an ungauged catchment in India, the case of Anas catchment [D]. Karlsruhe: Universität Karlsruhe, 2004

SLAMA C C, THEURER C, HENRIKSEN S W. Manual of photogrammetry[M]. 4th ed. Baton Rouge: American Society of Photogrammetry, 1980.

SLONECKER E T, SHAW D M, LILLESAND T M. Emerging legal and ethical issues in advanced remote sensing technology[J]. Photogrammetric Engineering & Remote Sensing, 1998, 64(6): 589-595.

SMITH P R, SARFATY R. Creating a strategic plan for configuration management using computer aided software engineering (CASE) tools[J]. Paper for National Doe/contractors & Facilities CAD/CAE Users Group, 1993, ml: 33-34.

SOLLER D R. Digital mapping techniques 99-workshop proceedings[R]. Madison: U. S. Geological Survey Open-File Report, 1999.

STAR J, ESTES J. Geographic information systems: an introduction [M]. Upper Saddle River: Prentice Hall, 1990.

SUCI A, BAYU R, AGUSTIN I, et al. Detection of spatial-temporal autocorrelation using multivariate Moran and

Lisa method on Dengue Hemorrhagic Fever (DHF) incidence, East Java, Indonesia[J]. European Journal of Scientific Research, 2011, 49(2): 279-285.

TIAN Y, GERKE M, VOSSELMAN G, et al. Automatic edge matching across an image sequence based on reliable points[C]// International Society for Photogrammetry and Remote Sensing (ISPRS). 21st Congress of the International Society for Photogrammetry and Remote Sensing, 2008.

TOMLIN C D. GIS and catograhic modeling[M]. Redlans: Esri Press, 2012.

TOMLINSON R F. Thinking about GIS: geographic information system planning for managers[M]. 3rd ed. Redlans: Esri Press, 2007.

TURBAN E, ARONSON J E. Decision support systems[M]. New York: Prentice Hall, 2005.

WEIBEL R, HELLER M. Digital terrain modelling[M]// Maguire D J. Geographical information systems: principles and applications. London: Longman, 1991

WEIBEL R. Concepts and experiments for the automation of relief generalisation [D]. Zurich: University of Zurich, 1989.

WEST M. Developing high quality data models[M]. San Francisco: Morgan Kaufmann Publishers Inc., 2011.

WHITE D. Relief modulated thematic mapping by computer[J]. The American Cartographer, 1985, 12(1): 62-67.

WU F. SimLand: A prototype to simulate land conversion through the integrated GIS and CA with AHP-derived transition rules[J]. International Journal of Geographical Information Science, 1998, 12(1): 63-82.

YAAKUP A, LUDIN A M, NAZRI A, et al. GIS in urban planning and management: Malaysian experience[J]. International Symposium & Exhibition on Geoinformation, 2005, 9(5): 27-29.

YOELI P. Analytical hill shading[J]. Surveying and Mapping, 1965, 25(4): 573-579.

YOELI P. The mechanisation of analytical hill shading[J]. Cartographic Journal, 1967, 4(2): 82-88.

ZHANG Q F, WANG L, WU F Q. GIS-based assessment of soil erosion at Nihe Gou Catchment[J]. Agricultural Sciences in China, 2008, 7(6): 101-105.

ZHOU Q M. Relief shading using digital elevation models[J]. Computer and Geosciences, 1992, 18(8): 1035-1045.